Computer Science, Technology and Applications

Anomaly Detection

Techniques and Applications

COMPUTER SCIENCE, TECHNOLOGY AND APPLICATIONS

Additional books and e-books in this series can be found on Nova's website under the Series tab.

COMPUTER SCIENCE, TECHNOLOGY AND APPLICATIONS

ANOMALY DETECTION

TECHNIQUES AND APPLICATIONS

SAIRA BANU
SHRIRAM RAGHUNATHAN
DINESH MAVALURU
AND
A. SYED MUSTAFA
EDITORS

Copyright © 2021 by Nova Science Publishers, Inc.

All rights reserved. No part of this book may be reproduced, stored in a retrieval system or transmitted in any form or by any means: electronic, electrostatic, magnetic, tape, mechanical photocopying, recording or otherwise without the written permission of the Publisher.

We have partnered with Copyright Clearance Center to make it easy for you to obtain permissions to reuse content from this publication. Simply navigate to this publication's page on Nova's website and locate the "Get Permission" button below the title description. This button is linked directly to the title's permission page on copyright.com. Alternatively, you can visit copyright.com and search by title, ISBN, or ISSN.

For further questions about using the service on copyright.com, please contact:
Copyright Clearance Center
Phone: +1-(978) 750-8400　　　　Fax: +1-(978) 750-4470　　　　E-mail: info@copyright.com.

NOTICE TO THE READER

The Publisher has taken reasonable care in the preparation of this book, but makes no expressed or implied warranty of any kind and assumes no responsibility for any errors or omissions. No liability is assumed for incidental or consequential damages in connection with or arising out of information contained in this book. The Publisher shall not be liable for any special, consequential, or exemplary damages resulting, in whole or in part, from the readers' use of, or reliance upon, this material. Any parts of this book based on government reports are so indicated and copyright is claimed for those parts to the extent applicable to compilations of such works.

Independent verification should be sought for any data, advice or recommendations contained in this book. In addition, no responsibility is assumed by the Publisher for any injury and/or damage to persons or property arising from any methods, products, instructions, ideas or otherwise contained in this publication.

This publication is designed to provide accurate and authoritative information with regard to the subject matter covered herein. It is sold with the clear understanding that the Publisher is not engaged in rendering legal or any other professional services. If legal or any other expert assistance is required, the services of a competent person should be sought. FROM A DECLARATION OF PARTICIPANTS JOINTLY ADOPTED BY A COMMITTEE OF THE AMERICAN BAR ASSOCIATION AND A COMMITTEE OF PUBLISHERS.

Additional color graphics may be available in the e-book version of this book.

Library of Congress Cataloging-in-Publication Data

ISBN: 978-1-53619-264-3

Published by Nova Science Publishers, Inc. † New York

CONTENTS

Preface		vii
Acknowledgment		ix
Chapter 1	Secured and Automated Key Establishment and Data Forwarding Scheme for the Internet of Things *N. V. Kousik, R. Arshath Raja, N. Yuvaraj and S. Anbu Chelian*	1
Chapter 2	A Study of Enhanced Anomaly Detection Techniques Using Evolutionary-Based Optimization for Improved Detection Accuracy *Vidhya Sathish and Sheik Abdul Khader*	19
Chapter 3	Anomaly Detection and Applications *Huichen Shu*	47
Chapter 4	An Evolutionary Study on SIoT (Social Internet of Things) *Dinesh Mavaluru and Jayabrabu Ramakrishnan*	77

Chapter 5	A Critical Study on Advanced Machine Learning Classification of Human Emotional State Recognition Using Facial Expressions *Jayabrabu Ramakrishnan and Dinesh Mavaluru*	93
Chapter 6	Anomaly Detection for Data Aggregation in Wireless Sensor Networks *Beski Prabaharan and Saira Banu*	139
Chapter 7	Algorithm for Real Time Anomalous User Detection from Call Detail Record *Saira Banu and Beski Prabaharan*	151
Chapter 8	Secured Transactions from the Anomaly User Using 2 Way SSL *Syed Mustafa and Mr. Madhivanan*	159
About the Editors		171
Index		173

PREFACE

Chapter 1 - This chapter presents the scheme for ensuring the security of data through the IoT sensors. LC-KES collaborative key management technique is used for guaranteeing the security of data transmitted to the servers. This chapter proposes the SAKE-MBAT-FNN and Modified BAT algorithm for the cryptographic action. The proposed algorithm increases the performance in terms of throughput and network life time.

Chapter 2 - This chapter discusses the anomaly based Intrusion Detection systems. The study related to utilization of evolutionary based hybrid approaches are presented and its successive rate in accurate extraction of 'malicious' patterns from existence when compared to other approaches are proven. This chapter also discusses the challenging research requirement which need to be fulfilled in the cyber research community.

Chapter 3 - This chapter introduces the structure of outlier detection algorithm. It also presents the techniques for detecting the outlier using the tools like python and Rapid miner. Proximity based, PCA, local outlier detection and high dimensional outlier detection techniques are explained in detail.

Chapter 4 - This chapter discusses the issues with respect to vulnerable data in IoT devices. It presents the CLA technique for detecting the vulnerable data and also highlights the challenges involved in handling those data.

Chapter 5 - This chapter determines the fundamentals for Facial Expression Recognition (FER) techniques. The various classifiers used for examining the facial expression recognition system are discussed and the results are compared. The best suited classifier for FER is identified.

Chapter 6: This chapter uses aggregation schemes for securing the data from anomaly users in the wireless sensor networks. This technique is used for both homogeneous and heterogeneous WSN.

Chapter 7 - This chapter discusses the methodology to find the anomalous users in telecommunication. The spam callers are detected using the information in the call detail record.

Chapter 8 - This chapter addresses the major challenge of securing the transactions from the anomaly user using the SSL and micro services. In this chapter. 2-way SSL is used for protecting the data using spring boot technology.

ACKNOWLEDGMENT

First and foremost, I thank the Almighty whose unbounded blessings and love have helped me in pursuing my life in the positive direction. I am grateful for my friends and colleagues for their continuous support and encouragement. I am also thankful to my family members for their best wishes and encouragement.

In: Anomaly Detection
Editors: Saira Banu et al.
ISBN: 978-1-53619-264-3
© 2021 Nova Science Publishers, Inc.

Chapter 1

SECURED AND AUTOMATED KEY ESTABLISHMENT AND DATA FORWARDING SCHEME FOR THE INTERNET OF THINGS

N. V. Kousik[1,*], R. Arshath Raja[2,**], N. Yuvaraj[3,ϵ] and S. Anbu Chelian[4,ϵϵ]

[1]Associate Professor,
School of Computing Science and Engineering,
Galgotias University, Greater Noida, Uttar Pradesh, India
[2]Senior Associate, Research and Development,
ICT Academy, Chennai, India
[3]Deputy Manager, Research and Development,
ICT Academy, Chennai, India
[4]Assistant Professor (Selection Grade),
Ramanujan Computing Centre, CEG Campus,
Anna University, Chennai, India

---

* Corresponding Author's Email: nvkousik@gmail.com.
** Corresponding Author's Email: arshathraja.ru@gmail.com.
ϵ Corresponding Author's Email: yraj1989@gmail.com.
ϵϵ Corresponding Author's Email: anbuchelianrcc@gmail.com.

ABSTRACT

Internet of things are most growing and popular field in the real world environment, which make life easy and seamless for people. IOT utilizes the sensors and internet to monitor and inform the day-to-day activities of people, thus helping them to take important actions on time. Sensor devices fixed on the appliances monitor and gather the information, which are then sent to the servers through intermediate proxies. Involvement of malicious nodes might lead to data stealing, which is resolved in the existing research work by using the methodology of Lightweight Collaborative-Key Establishment Scheme (LC-KES) for ensuring secure data transmission to the servers. This work ensures security by doing collaborative key management between sensor, proxies and server. Here, proxies are assumed to be less resource-constrained and it is selected randomly for performing cryptographic function. However, in the real world, proxies might lead to failure due to resource drainage as these are wireless devices that are not foolproof. In the proposed research methodology, secured cryptography is ensured by selecting the optimal proxies, which can withstand until completion of cryptographic action. In the proposed research methodology, novel framework is proposed in the form of Secured and Automated Key Establishment using Modified BAT and Fuzzy Neural Network (SAKE-MBAT-FNN) technique. In the proposed research work, MBAT algorithm is used for optimal selection of the proxy server in terms of QoS constraints, where the cryptography would be performed. And then automated key generation is performed using FNN technique so that computational load due to more memory consumption can be avoided. The simulation result shows that the SAKE-MBAT-FNN archives improved security results than the existing methods.

Keywords: Secured key establishment, resource constraints, optimal proxies, adaptive handling and computation overhead avoidance

1.1. INTRODUCTION

The Internet of Things (Internet of Things (Iot) is defined as the system of computing devices with unique identifiers that are connected with each other [1]. Each device in IoT is assigned with a specific IP address, which has the ability to transfer the network data [2]. There exists several applications of IoT that includes micro-electromechanical systems (MEMS), wireless technologies, etc. IoT allows the analysis using unstructured machine-generated data to drive improvements [4].

The creation of a secure end-to-end channel between remote entities focuses on one of the IoT technologies, namely Wireless Sensor Networks (WSNs). The WSN enable the devices to communicate with other entities about their surroundings. The sensor network elements must therefore be allowed to connect with other entities via the internet. The issues concerning data flow security are, however, not insignificant. Sensor nodes are usually restricted devices that have very limited computing power and may be too heavy too in case of negotiating a session key with other IoT entities in key management mechanism [5]. This work, therefore, focuses on the implementation of secure key management systems that can be applied to Internet sensor networks often employed in actual internet scenarios. However, in our study, we assess the key management focused on negotiations on link-layer keys between the neighboring nodes. Obviously, the internet scenario did not take these mechanisms into account. However, the present research corpus in this specific field is sufficiently large enough to warrant a study of its applicability.

The overall organization of this research chapter is given as follows: In section 2, existing methods implemented for solving the key management problems are discussed. In section 3, detailed discussion is given as per the proposed work. In section 4, experimental evaluation is given. Finally, in section 5, conclusion is discussed with possible directions of future work.

1.2. RELATED WORKS

In recent years, it (what does that 'it' mean? Please explain) was a highly active field for research to develop Key Management Systems (KMS) to set the link layer keys in WSN nodes [6]. In the key pool framework, KMS framework is one of the most important frameworks proposed so far, where the key role is played by key pool paradigm. This framework's basic scheme is very simple [7]. Firstly, a key pool creates the network designer, namely a large set of secret keys that are pre-calculated. Secondly, each sensor node is assigned a single key chain that is to say a small subset of the key pool keys before deployment. Third, the nodes exchange their key identification numbers from the key chains after network deployment in an attempt to find a common secret key shared. If the two Nodes fail to share similar keys, a key route or secure routing path is not found between them to negotiate a pair-wise key. The 'pool' concept can further be applied on various frameworks of other KMS. The mathematical framework on KMS use mathematical functions to calculate link layer keys [8]. The nodes can also negotiate their keys right after the deployment of WSN with their close neighbour. The negotiation framework includes all protocols generating their keys by mutual agreement, and usually assumes that there are few or no threats to the integrity of a WSN network in the very early stages of life. In addition, all KMSs considering a network-a large-scale [12] and protocols that focus on network organization are included in this particular framework into dynamic or static clusters. Notice that the protocols of this framework can be optimized in order to ensure the authenticity of pairs at all levels of the network development [14]. Finally, each KMS protocol can be optimized in several ways from the above frameworks. For example, knowledge of the locations of final sensor node in WSN can be used for reduction of overhead communication and memory consumption. Additional optimizations are aimed at reducing the size of messages and total number of messages that are used for finding the common key. In order to enhance certain properties like extensibility and resilience, there are some improvements that can be made to the underlying KMS structure.

1.3. SECURED AND AUTOMATED KEY ESTABLISHMENT MANAGEMENT

The light weight collaborative key establishment scheme (LC - KES) system for the secure transmission of data to servers is currently being developed. They are based on the reliable delivery of all xi secret fragments in order to reconstitute the secret key of the source on the destination node. The information for a server is incomplete due to single missing proxy message and protocol exchange may fail during exchange. If proxies act as honest and reliable actors, the reliability of such proxies is not guaranteed even if the scenario involves the proxies that are trustworthy for resource restricted nodes.

Therefore, in the event of an unavailability or non-cooperative proxy behaviour, a transmission optionally preceded by a new proxy assignment may be required. An additional latency will however be present in the existing system.

The main problem that can be found on the existing system is sensor nodes that would select the number of proxies randomly to send their messages, where the cryptographic action would be carried out. The existing work assumes that those proxy servers have less resource constraints and there will not be any failure. However, in real world scenario, all the proxy servers are wireless devices, which are resource-constrained in nature. Thus the random selection of proxies would lead to security violations by not completing the cryptographic functions completely.

The SAKE-MBAT-FNN resolves this issue by the introduction of a novel framework using Secured and Automated Key Establishment using Modified BAT and Fuzzy Neural Network (SAKE-MBAT-FNN) technique. In the proposed research work, modified BAT algorithm is used for optimal selection of the proxy server in terms of QoS constraints, where the cryptographic actions would be performed. And then automated key handling is performed using FNN technique so that higher security can be achieved. The architecture of the SAKE-MBAT-FNNology is given in figure 1.

In the figure 1, processing flow of the SAKE-MBAT-FNN is shown, where the secured transmission is made by using efficient key establishment management procedure. The secured transmission is achieved by selecting the more optimal proxy servers in the proposed research methodology.

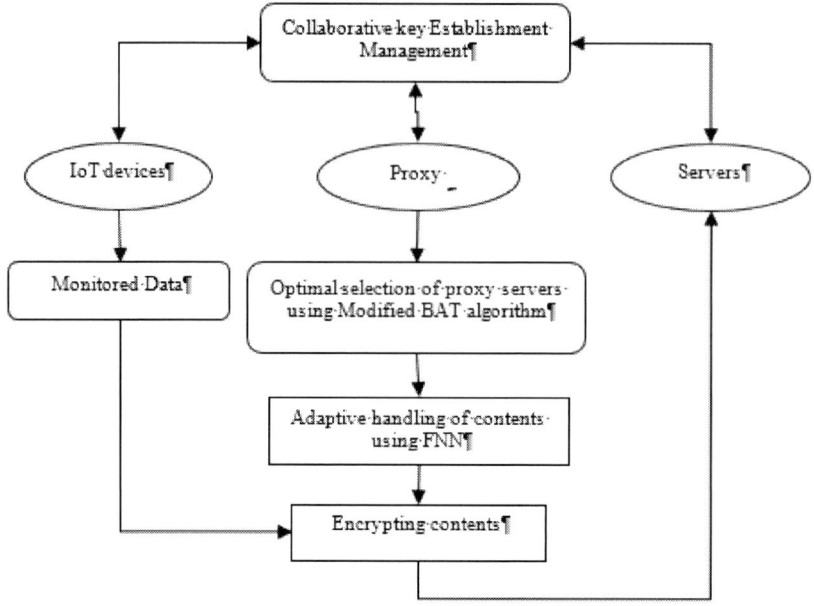

Figure 1. Overall processing flow of SAKE-MBAT-FNN.

1.4. Network Model

The model of the network considers an IoT infrastructure that connects heterogeneous nodes with various computer power and energy resources capabilities. The proposed work classifies the nodes into three types based on the services: proxy nodes, highly-resourceful nodes and unconstrained nodes.

The resource-limited sensor node A, carefully selects the $P_1,...,P_n$ proxies to support its key exchanges on the basis of the reputation and resources of the network nodes. Our approach requires that during the main exchange, these nodes process messages for the resource - constricted node. Therefore, questions of authorization and authentication occur on the proxy side, as these nodes must be given proof of representativeness. This could also be the public key of the proxy associated with the right "authority for signing on behalf of A" signed with private key of the source and supplied "offline" to the proxy. This evidence could be the official certificate. However, it is possible to mislead the use of long - term certificates of authorization. Proxy, when the cryptographic actions would be carried out.

Thus the selection of optimal proxy servers plays a more important role, which is done in the proposed research methodology using Modified BAT algorithm.

1.5. OPTIMAL PROXY SERVER SELECTION USING MODIFIED BAT ALGORITHM

Proxy server plays a more important role in the proposed research methodology, which needs to be selected more carefully to fulfil the data transmission requirements. By selecting the proxy servers with complete resources cryptographic functions can be performed without any interruption or failure. The constraints that are considered for the optimal selection of proxy are bandwidth and energy consumption which is calculated as follows:

Available Bandwidth (BW): The bandwidth (BW) is defined as the link available between source and destination node. The bandwidth is determined with Eqn. 1.

$$BW = \alpha BW_L + (1 - \alpha) \frac{T_{idle}}{T_p} B_{channel}$$

where, α- weight factor, BW_L- available bandwidth, T_{idle}- idle time, t_p- time interval period and $B_{Channel}$- channel capacity.

Available Power (P): The available node power is expressed as follows:

$$P = P_{Total} - E_{consumed}$$

where P_{Total} - total node energy and hence the $P_{Consumed}$ is estimated as,

$$P_{consumed} = \frac{P_{Threshold} d^n}{K}$$

where $P_{Threshold}$- predefined threshold power, n - path loss exponent, d - distance between two sensor nodes and K - predefined constant.

1.5.1. Modified Bat Algorithm

The Bat algorithm process flow is given in the following sub-section:
a) Initialization of bat population:
In the given population, the food sources quality is assessed as:

$$x_{ij} = x_{min} + \varphi (x_{max} - x_{min}) \tag{1}$$

where

i = 1,2,…..,N,
j = 1,2,….d,

x_{max} is the upper bound and x_{min} is the lower bound
b) Generation of frequency, velocity and new solutions:
Bats moving with a velocity vi is highly influenced by the frequency f in the search space and its position xi is estimated as

$$f_i = f_{min} + \beta (f_{max} - f_{min}) \tag{2}$$

$$v_i^t = v_i^{t-1} + (x_i^t - x_*)f_i \tag{3}$$

$$x_i^t = x_i^{t-1} + v_i^t \tag{4}$$

where f_i- frequency of ith bat, f_{min} is the minimum frequency and f_{max} is the maximum frequency, b - generated randomly value, x^*- global best solution for the location of bats among N bats at t.

c) Local search capability:

A structure is created to improve the search capacity so that the bats improve the solution that is optimal to the solution algorithm.

$$x_{new} = x_{old} + \varepsilon \overline{A}^T \tag{5}$$

where x_{old} is a high quality solution chosen by some mechanism (e.g., roulette wheel), $\overline{A}^T$ is average loudness value of all bats at tth time step and ε is a randomly generated value ranging from -1 to 1.

d) Loudness and pulse emission rate:

In the case of bat approaching its target, namely the prey, the emission of the pulse and loudness A are updated.

The reduction in loudness A is increased with pulse emission r in relation to Eqs. 6 and 7, each.

$$A_i^{t+1} = \alpha A_i^t \tag{6}$$

$$r_i^{t+1} = r_i^0 (1 - e^{\gamma t}) \tag{7}$$

where γ and α- constraints, r_i^0- pulse emission rate. Weight modification is provided below.

$$v_i^t = \omega \left(v_i^{t-1}\right) + \left(x_i^t - x_*\right) f_i \tag{8}$$

where w - weight inertia factor with local and global search intensity with velocity v.

ALGORITHM: Modified BAT Algorithm
Step 1: Initialize bat population x_i and velocity v_i
Step 2: Define frequency f_i
Step3: Initialize pulse emission rate r and loudness A
Step 4: repeat
Step 5: Generate new solutions by adjusting frequency and updating velocity and location by Equations 2 to Equations 4

Step 6: if rand>r_i then
Step 7: Select a solution among best solutions
Step 8: Generate new local solution around selected best solution
Step 9: end
Step 10: Generate new solution by flying randomly
Step 11: if rand<A_i and f (x_i)<f (x_*) then
Step 12: Accept the new solution
Step 13: Decrease A_i, increase r_i, by Equations 6 and 7
Step 14: end
Step 15: Rank the bats and find the current best x_*
Step 16: until termination criteria is met;
Step 17: Post process results and visualization

1.6. COLLOBORATIVE KEY HANDLING MANAGEMENT

The approach being distributed is based on a (k, n) threshold scheme, whereas k polynomial shares are k enough to reconstruct a public DH source key via the Lagrange polynomial interpolation technique, while n proxies get a polynomial share rather than a partition element. In cryptography, Shamir's secret sharing schemes initially use Lagrange polynomials. Given a polynomial function f of degree k-1 is expressed as: $f(x) = q_0 + q_1 x + \ldots + q_{k-1} x^{k-1}$ with $q_1, q_2, \ldots, q_{k-1}$ that is considered as uniform, random and independent coefficients and $q_0 = a$. A Lagrange formula is applied to retrieve the polynomial f, which is given as follows:

$$f(x) = \sum_{i=1}^{k} \left(f(i) \times \prod_{j=1, j \neq i}^{k} \frac{x-j}{i-j} \right) \qquad (9)$$

From (9), the secret exponent is computed for any k subset values of f(x):

$$a = f(0) = \sum_{i=1}^{k} \left(f(i) \times \prod_{j=1, j \neq i}^{k} \frac{-j}{i-j} \right) \qquad (10)$$

Upon the reception of a subset P of k values transmitted by the proxies, the server starts by computing the c_i coefficients as follows:

$$C_i = \prod_{i \in P, j \neq i} \frac{-j}{i-j} \qquad (11)$$

Then, B uses Lagrange formula to compute the public key of source DH, DH_I, which is given as

$$\prod_{i \in P} \left(g^{f(i)}\right)^{c_i} \bmod p = g^{\sum_{i \in P} f(i) \times c_i} \bmod p = g^{f(0)} \bmod p = g^a \bmod p \quad (12)$$

In order to prepare the computation of the DH session key at the source side, B starts calculating for each proxy P_i (iϵ P) the value $B_i = g^{b.c_i} \bmod p$ (c_i being the ith coefficient calculated in the previous phase). P_i is unable to compute the coefficient ci since it has no knowledge about the subset P of concrete participating proxies.

Having received this value, each proxy P_i uses its share f(i) of the source's private exponent to compute $K_i = B_i^{f(i)} = g^{b.c_i.f(i)}$. Each proxy then delivers this computed value to the source A. Upon reception of these k values, the source computes the DH session key KDH as follows:

$$K_{DH} = \prod_{i \in P} g^{bf(i)c_i} \bmod p = g^{b \sum_{i \in P} f(i) c_i} \bmod p = g^{ab} \bmod p \quad (13)$$

By using the threshold technique, the source will perform further computer operations during the initial phase for the calculation of polynomial n values, which is sent to n proxies. The computation cost is better estimated if another way to write f(x) is taken into account:

$$f(x) = (\ldots((q_{k-1}x + q_{k-2}).x + q_{k-3})x + \cdots).x + q_0 \qquad (14)$$

According to this expression, A performs for each computation of f(i): (k - 1) multiplications between a scalar and a large number and (k - 1) summations of two large numbers. It should be noted that k and n are smaller than the number of secure relationships the source can maintain.

The polynomial coefficients, on the other hand, are as big as the DH source's private key.

1.7. KEY HANDLING PROCEDURE

There might be an issue occurring when there is a presence of malicious nodes which can steal the keys that are generated. Both networks share their results and when both networks are synchronized, the key exists between the two parties with final learned weights. The neural synchronization security is jeopardized if an attacker can synchronize in the training with either of the two parties. But if the algorithm is robust and the keys are long, random and unpredictable, then the crypto - system security remains strong. The keys are then deployed on a cryptography system using Data Encryption Standard encryption decryption encryption.

Data Gathering and Pre-processing: The collection of data included the assembly of all the data used for the formation of the neural network ARTMAP.

Training and Testing Loops: Only in the field of architecture was the algorithm trained.

Network Deployment: In the MATLAB program, the neural network was used as a module.

Independent Testing and Verification: The system was tested for verification when all the limitations were fulfilled and all activities documented.

1.8. EXPERIMENTAL RESULTS

In this section, the performance metrics are evaluated with SAKE-MBAT-FNN. The performance metrics are such as packet delivery ratio, end-to-end delay, throughput and network lifetime is evaluated by using existing LC-KES method and proposed SAKE-MBAT-FNN method.

Proposed SAKE-MBAT-FNN method is evaluated using MATLAB 13. The proposed SAKE-MBAT-FNN method is compared with existing Threshold Distributed TLS (TDTLS), DTLS.

Table 1. Simulation parameters

Area	100*100
Initial energy	0.5J
Nodes number	100
Packet size	4000bits
BS location	(50,50)
E	50nJ/bit
ε_{mp}	0.0013pJ/bit
ε_{fs}	10pJ/bit/m2

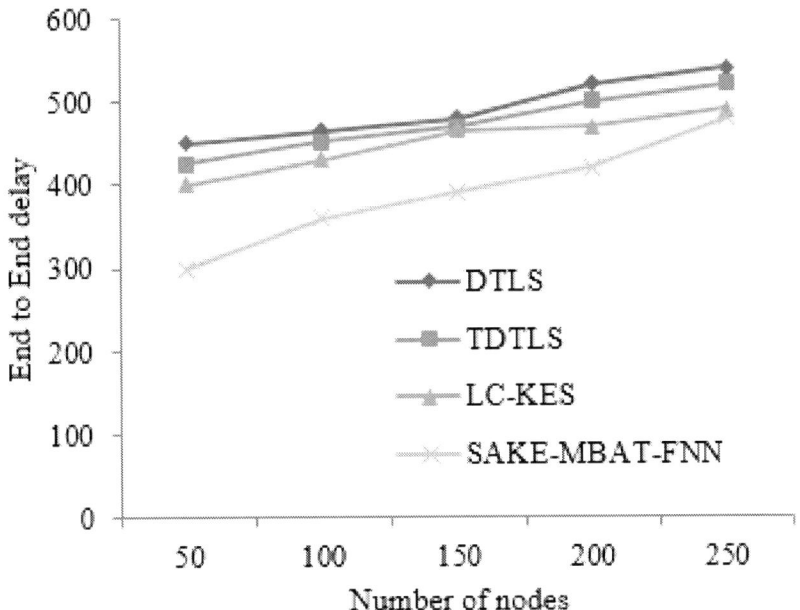

Figure 2. End to End delay comparison.

The simulation parameters are given in table 1.

In Figure 2, we observe the comparison of SAKE-MBAT-FNN with existing methods in terms of end-to-end delay metric.

In SAKE-MBAT-FNN, the end-to-end delay value is reduced significantly by using the efficient detection of SAKE-MBAT-FNN method.

From figure 3, we can observe that the comparison of SAKE-MBAT-FNN with existing methods in terms of network lifetime metric.

In SAKE-MBAT-FNN, the network lifetime value is increased significantly by using the SAKE-MBAT-FNN method. Thus it shows that the SAKE-MBAT-FNN performs efficient detection using the SAKE-MBAT-FNN.

Figure 4 shows the results of existing and SAKE-MBAT-FNN in terms of throughput metric.

In SAKE-MBAT-FNN, the throughput is improved by the use of SAKE-MBAT-FNN method.

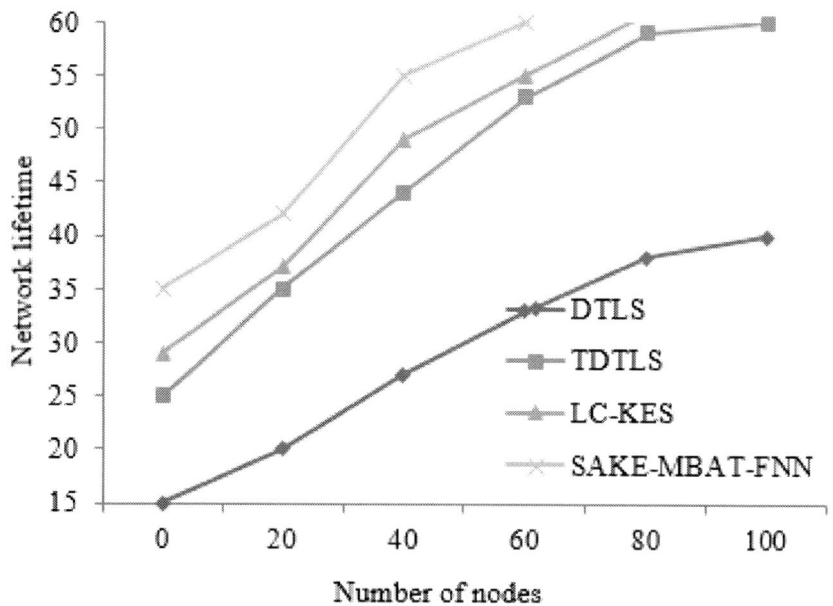

Figure 3. Network Lifetime comparison.

Secured and Automated Key Establishment ... 15

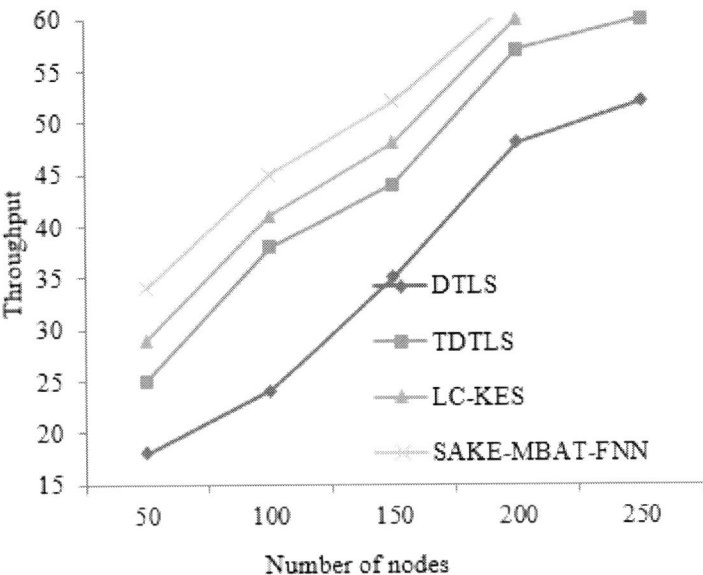

Figure 4. Throughput comparison.

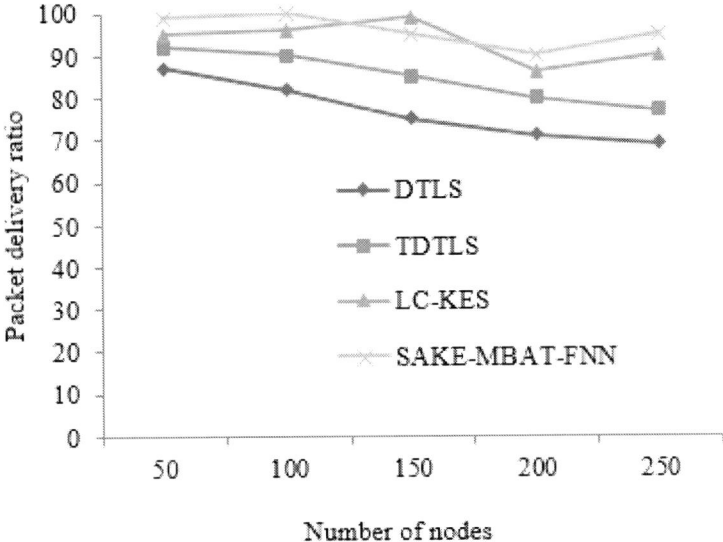

Figure 5. Packet delivery ratio comparison.

From the above figure we can observe that the comparison of existing and SAKE-MBAT-FNN in terms of packet delivery ratio.

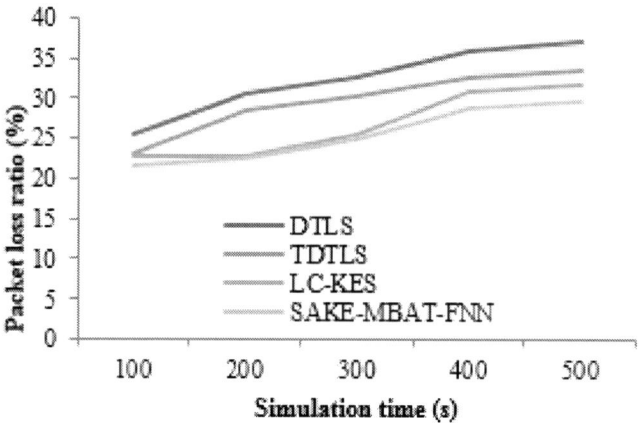

Figure 6. Packet loss ratio comparison.

In SAKE-MBAT-FNN, the PDR is increased significantly with reduced number of nodes.

In Figure 6, when the simulation time is 500 seconds, the SAKE-MBAT-FNN achieves a loss ratio of 28.59%, which is 9.59%, 4.95 and 1.04% less than LC-KES, TDTLS and DTLS respectively. Thus it is clear that the SAKE-MBAT-FNN offers improved performance than other methods.

CONCLUSION

In this work, Secured and Automated Key Establishment using Modified BAT and Fuzzy Neural Network (SAKE-MBAT-FNN) technique aim to establish the secured key handling procedure by selecting the proxy servers optimally and thus ensuring the secured data transmission. Optimal proxy server selection is done with the consideration of the parameters energy and bandwidth, thus the continuous and reliable delivery of data is made sure with the help of modified BAT algorithm. The secured key generation is ensured by adapting the fuzzy neural network scheme, which can produce the keys that are difficult to predict. The overall analysis is conducted in the matlab simulation environment by implementing and testing under different performance parameter values.

REFERENCES

[1] Jun, Zhang, Simplot-Ryl, D., Bisdikian, C. and Mouftah, H. T. "The internet of things". *IEEE Commun. Mag*, 49, no. 11 (2011): 30 - 31.

[2] Da Xu, Li, Wu He and Shancang Li. "Internet of things in industries: A survey". *IEEE Transactions on Industrial Informatics,* 10, no. 4 (2014): 2233 - 2243.

[3] Al-Fuqaha, Ala, Mohsen Guizani, Mehdi Mohammadi, Mohammed Aledhari and Moussa Ayyash. "Internet of things: A survey on enabling technologies, protocols, and applications". *IEEE Communications Surveys and Tutorials,* 17, no. 4 (2015): 2347 - 2376.

[4] Singh, Dhananjay, Gaurav Tripathi and Jara, Antonio J. "A survey of Internet-of-Things: Future vision, architecture, challenges and services". In Internet of things (WF-IoT), 2014. *IEEE World Forum on,* pp. 287 - 292. IEEE, 2014.

[5] Roman, Rodrigo, Alcaraz, Cristina, Lopez, Javier and Sklavos, Nicolas. "Key management systems for sensor networks in the context of the Internet of Things". *Computers and Electrical Engineering,* 37, no. 2 (2011): 147 - 159.

[6] Barreto, M. A. S. Jr., P. S., Margi, C. B., Carvalho, T. C. A Survey on Key Management Mechanisms for Distributed Wireless Sensor Networks, *Computer Networks,* 54 (15) (2010) 2591 - 2612.

[7] Eschenauer, L., Gligor, V. D. A Key-Management Scheme for Distributed Sensor Networks, in: *9th ACM conference on Computer and Communications Security (CCS 2002),* 2002, pp. 41 - 47.

[8] Du, W., Deng, J., Han, Y. S., Varshney, P., Katz, J., Khalili, A. A Pairwise Key Pre-distribution Scheme for Wireless Sensor Networks, *ACM Transactions on Information and System Security (TISSEC),* 8 (2) (2005) 228 - 258.

[9] Camtepe, S. A., Yener, B. Combinatorial Design of Key Distribution Mechanisms for Wireless Sensor Networks, *IEEE/ACM Transactions on Networking,* 15 (2) (2007) 346 - 358.

[10] Liu, D., Ning, P., Li, R. Establishing Pairwise Keys in Distributed Sensor Networks, *ACM Transactions on Information and System Security,* 8 (1) (2005) 41 - 77.
[11] Anderson, R., Chan, H., Perrig, A. Key infection: Smart Trust for Smart Dust, in: *Proceedings of the 12th IEEE International Conference on Network Protocols (ICNP),* 2004, pp. 206 - 215.
[12] Lai, B., Kim, S., Verbauwhede, I. Scalable Session Key Construction Protocol for Wireless Sensor Networks, in: *IEEE Workshop on Large Scale Real-Time and Embedded Systems (LARTES 2002),* 2002.
[13] Panja, B., Madria, S., Bhargava, B. Energy and Communication Efficient Group Key Management Protocol for Hierarchical Sensor Networks, in: *Proceedings of the IEEE International Conference on Sensor Networks, Ubiquitous, and Trustworthy Computing (SUTC'06),* Vol. 1, 2006, pp.1 - 8.
[14] Seshadri, A., Luk, M., Perrig, A. SAKE: Software Attestation for Key Establishment in Sensor Networks, *Ad Hoc Networks* In Press, Corrected Proof.
[15] Yang, X.-S. A new metaheuristic bat-inspired algorithm, in: Gonzlez, J., Pelta, D., Cruz, C., Terrazas, G., Krasnogor, N. (Eds.), *Nature Inspired Cooperative Strategies for Optimization (NICSO 2010),* Vol. 284 of Studies in Computational Intelligence, Springer Berlin Heidelberg, 2010, pp. 65 - 74.
[16] Fenton, M. Bat natural history and echolocation, in: Brigham, R., Elisabeth, K., Gareth, J., Stuart, P., Herman, A. (Eds.), Bat Echolocation Research tools, techniques and analysis, *Bat Conservation International,* 2004, pp. 2 - 6.
[17] Kennedy, J., Eberhart, R. Particle swarm optimization, in: Neural Networks, 1995. Proceedings, *IEEE International Conference on,* Vol. 4, 1995, pp. 1942 - 1948 vol. 4.
[18] Shi, Y., Eberhart, R. A modified particle swarm optimizer, in: Evolutionary Computation Proceedings, 1998. IEEE World Congress on Computational Intelligence. *The 1998 IEEE International Conference on,* 1998, pp. 69 - 73.

In: Anomaly Detection
Editors: Saira Banu et al.
ISBN: 978-1-53619-264-3
© 2021 Nova Science Publishers, Inc.

Chapter 2

A STUDY OF ENHANCED ANOMALY DETECTION TECHNIQUES USING EVOLUTIONARY-BASED OPTIMIZATION FOR IMPROVED DETECTION ACCURACY

Vidhya Sathish[1], PhD and Sheik Abdul Khader[2], PhD
[1]Assistant Professor, Shri Krishnaswamy College for Women, Chennai
[2]Professor, B. S. Abdur Rahman Cresent Institute of Science and Technology, Vandalur, Chennai

ABSTRACT

The obsession of maintaining cyber security seems to be a research challenge for the internet community. Trapping of zero-day exploits from the normal traffic pattern has been highlighted as a major research challenge. The zero-day exploits are groomed in terms of virus, Trojans, etc. There are two major requirements for the designing of accurate extraction of 'normal' and 'malicious' pattern analysis. They are notified as High detection accuracy and Low false positive rate. The detection accuracy is necessary to be higher for reducing the declassification of 'malicious' traffic pattern as 'normal' traffic pattern, i.e., reducing of false leads. Based on these criteria, projection of research is on progress over the past two decades. But due to the evolution of zero-day exploit, the research challenge is still in existence. Generally, detection methodologies had been categorized in two ways. They are known to be Signature-based Intrusion Detection systems and Anomaly-based Intrusion Detection Systems. The Signature-based Intrusion Detection systems focus with known patterns, which are already trained by system itself. analyses of successive rate of these detections show its extremity in the extraction of trained signature patterns. But it limits upon certain extent, i.e., fails to find zero-day exploits. In such a case, the system gets collapsed, which may lead to huge machinery loses.

To overcome this, anomaly-based Intrusion Detection systems were developed. Commonly, for designing the detection techniques, classifier-based approaches are evolved. These types of approaches are composed with data mining-based and machine learning-based design techniques, which have been widely utilized in both Signature-based detection techniques and anomaly-based detection techniques. The hybrid approach of data mining and machine learning-based techniques have proven their success rate in the extraction of anomalies in a significant way when compared to individual approach. In terms of designing the anomaly detection technique harder, evolutionary-based technique was developed. The goal of this technique has focused on the pheromone behavior of species, mammals or ants during hunting of their prey. Similarly, artificial intelligence-based techniques may prompt to reach the higher level of extraction of 'normal' and 'malicious' traces by reducing its false leads within the minimal learning time. The objective of this chapter is to discuss the anomaly-based Intrusion Detection systems elaborately and also present the study-related to utilization of evolutionary-based hybrid approaches, proving its success rate in accurate extraction of 'malicious' patterns from existence when compared to other approaches. This chapter also discusses the challenging research requirement which needs to be fulfilled in the cyber research community.

2.1. INTRODUCTION

The extraction of intrusions residing over the cybernetics seems to be a major manifesto for the research community [1]. Generally, intrusions have been classified into four different types of patterns, which turn into damage to systems at any cause. They are known to be Type1: Denial-of-Service attacks that mock the system performance from ideal to busy state.

Type 2: Probing attacks that initiate network scanning with a malicious intent to threat the system information. Type 3 and Type 4; User-to-Root and Root-to-Local attacks target the systems to be attacked from one or more number of systems to loot the information through authenticated user. There are types of intrusion trace known as virus, Trojans, DDoS, bot, etc., malware are that had been endeavoring with neither the intent to affe

ct the system, either internally or externally [2]. Research to analyze these intrusions traces begin with Signature-based Intrusion Detection systems [3]. The behavior of this detection systems, analyze the known signature patterns, which are already trained by the system. The successive rate of these detection systems has proven their efficiency in representing the known signature patterns and on the other hand prompt the researchers to study, design and analyze the anomaly-based Intrusion Detection systems in descriptive manner [4]. Detection over anomalies was identified to be a major challenging task due to evolution of zero-day exploits in today's environment. The research has been progressing from the past two decades, but still existence of intrusions turned up to be a complicated task [5]. Commonly, detection systems have been classified into two ways: They are known to be Signature-based Intrusion Detection systems and Anomaly-based Intrusion Detection systems. Both the systems are highly imposed by data mining and machine learning-based techniques [6].

Anomaly-based detection performance seems to be better course of action in identifying the intrusions when compared to Signature-based Intrusion detection systems. But the setback of Anomaly-based detection systems was found to be difficult in reducing the false alarm rate or false positives [7]. There are two major extensive requirements that needed to be considered for designing the efficient intrusion detection system: They are High Detection Accuracy and Low False alarm Rate [8]. The existing techniques though achieved its detection rate in a prominent way, seem to be imbalanced between these two requirements [9]. In earlier days, Anomaly-based detection techniques were highlighted with classifier-based detection techniques. The classifier-based detection approaches are composed of data mining and machine learning techniques.

The characteristics of data mining-based techniques focus on the prediction of unknown facts based on the analysis of processing step for knowledge discovery in databases. Machine learning-based techniques on the other hand focus on the prediction of known facts. Generally, both data mining and machine learning techniques are deployed with common techniques [10] with intent to improve the learner's accuracy. The success rates of these classifier-based approaches have proven their sufficiency in extraction of anomalies and also have showed their betterment when compared to signature-based detection approaches. In order to make anomaly detection techniques harder, evolutionary-based detection techniques were developed. The major uniqueness of these techniques [12] is focused with pheromone behavior of real species, mammals, ants, etc., whilst hunting a prey. Moreover, the evolutionary-based detection approaches show their eminent performance in reducing their false leads when compared to classifier-based techniques [13]. In recent days, hybrid approaches of classifier and evolutionary-based detection techniques have proven their efficiency in designing the efficient detection system when compared to one-hand approach.

2.2. LITERATURE ANALYSIS

In common term, Anomaly-based Intrusion Detection systems are classified into two categories. They are known to be classifier-based detection approaches and evolutionary-based detection techniques. This section will elaborately analyze the varied detection structures for designing the eminent intrusion detection system. The literature analysis also discusses about the overall limitations along with the detection accuracy and false alarm rate attained. The Figure 1 illustrates the methodologies utilized at different stages.

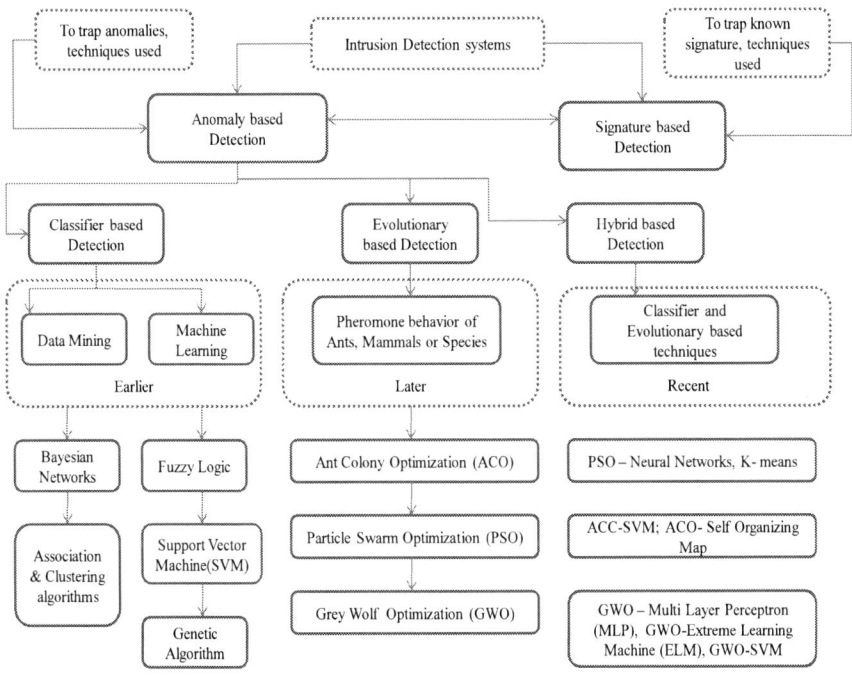

Figure 1. Classification of Intrusion Detection Systems.

There are the challenges highlighted by Intrusion Detection Systems in existence [14]. First, the issue related to reduction of false alarms. This issue has been notified as an essential requirement for designing the eminent Intrusion Detection system. Some of the computational tasks have been developed to attain this, i.e., the minimizing the false positives. But there is a lack of accuracy that occurred, while missing the real-time traces. So the design of the eminent Intrusion Detection system is needed to attain an improved detection accuracy by minimizing its false leads.

Second, the mobilization of attacks may disrupt the system functioning in several ways from authenticated user information such as looting the user's information, flooding of systems to crash the services and make compromise on system integrity from the negligence of real-time intrusion traces. Third, the presence of intrusion instances may be assembled neither internally nor externally with intent to threat the authenticated information using vulnerability exploitation techniques. Fourth, the theory of hybrid approach in analyzing the extraction of 'normal' and 'malicious' instances should be trouble-free quest. Therefore, Intrusion Detection systems need to work on the platform independently. Finally, extracting the 'malicious' pattern from the 'normal' traffic pattern analysis seems to be difficult and it still remains as open challenge in cyber research community [15]. Therefore, clustering over constraints of 'normal' and 'malicious' pattern analysis need improvement for better attainment.

2.3. CLASSIFIER-BASED APPROACHES

The design of diversified classified-based Intrusion Detection systems plays a keen role to prompt the eminent detection model. The objective of these detection systems focused to attain High Detection accuracy Rate (HDR) and Low False alarm Rate (LFR). The study of designing and analyzing the extraction of malicious pattern over Host-based network traffic [16] seems to be a resilient stint. It also discusses about the open research challenge and need to take necessary actions as a future trend. The influence characteristics of Intrusion Detection systems have focuses to set forth as a system of operating to investigate, dissect and detect the anomalies that existed [17]. The broad category of Anomaly-based detection system is the object of focus on optimal feature extraction.

The active response of analytical tools has the ability to disrupt intrusive behavior and this characteristic helps the classification technique in getting delineated as computationally inexpensive, diversified methodology for data processing and categorized into accurate classes. Generally, the consequent functioning of feature extraction make the classifier techniques evolved with two phases during the execution of work: the training phase and the testing phase. In the training phase, a heuristic value from the trained intrusion detection data is checked. The extracted feature from the heuristic value is compared with real traffic patterns in the testing phase for prediction. The performance of predicting the nearby value notified as data classification accuracy. The classification techniques have generally evolved as the threshold model to analyze the maximum separation between 'normal' and 'malicious' pattern behavior.

Table 1. Description of KDDCUP99 data

ATTACK TYPE	ATTACK NAME	ATTRIBUTE SEQUENCE	ATTRIBUTE SEQUENCE
Type 1 : DOS attacks	Back Land Neptune POD Smurf Teardrop	1. duration 2. protocol_Type 3. service 4. flag 5. src_bytes 6. dst_bytes 7. land	26. srv_serror_rate 27. rerror_rate 28. srv_rerror_rate 29. same_srv_rate 30. diff_srv_rate 31. srv_diff_host_rate 32. dst_host_count
Type 2 : User-to-Root attacks	Buffer overflow Load Module Perl Rootkit	8. wrong_fragment 9. urgent 10. hot 11. num_failed_logins 12. logged_in 13. num_compromised	33. dst_host_srv_count 34. dst_host_same_srv_rate 35. dst_host_diff_srv_rate 36. dst_host_same_src_port_rate 37. dst_host_srv_diff_host_rate 38. dst_host_serror_rate
Type 3 : Probing attacks	IPSweep Nmap Port Sweep Satan	14. root_shell 15. su_attempted 16. num_root 17. num_file_creations 18. num_shells 19. num_access_files 20. num_outbound_cmds	39. dst_host_srv_serror_rate 40. dst_host_rerror_rate 41. dst_host_srv_rerror_rate 42. label.
Type 4 : Root-to-Local attacks	FTP_write Guess_Passwd Imap Multihop Phf Spy	21. is_host_login 22. is_guest_login 23. count 24. srv_count 25. serror_rate	

The classifier-based optimized techniques endorsed with diversified analysis were conducted [18]. This project the network-based clustering analysis designed for protocol independent structure and the objective of this framework to implement the correlation analysis [19] in order to identify the hosts running between 'normal' and 'malicious' behavior. The distinguished clustering-based detection techniques had been reformed to identify the group of items with similar characteristics from the large set of composed elements. It discussed [20] about the diversified data mining techniques to enhance the Intrusion Detection analysis such as classification, varied clustering approaches like K-means clustering, Y-means clustering, Fuzzy C-means clustering and Association rules. In this technique, clustering approach is labelled as the best option when compared to other two approaches. Honeypot-based detection approaches [21] have proven their efficiency in extracting malicious activities. The perversion of these approaches is broadcast as a reactive-based detection approaches for network sniffing. This may lead to occurrence of more false positives. The knowledge-based techniques were developed to show its success rate in reduction of false positives [22].

Table 2. Attack name with State of interaction

ATTACK NAME	STATE OF INTERACTION	ATTACK NAME	STATE OF INTERACTION
1. back	client uniform resource locator with backslashes	11. IPsweep	attack exists by sweeping the network
2. land	sending false packet to source and destination	12. Nmap	examining the network fingerprinting for vulnerability purpose
3. neptune	malicious network pattern make the system resource to be busy using false IP addresses	13. Portsweep	make the multihosts to share the common network
4. pod	make the system to crash over the network	14. satan	leads to mis-configuration
5. smurf	generates the spoofed source IP addresses	15. ftp_write	Perform as 'man in the middle' attack.
6. teardrop	cause the packet to overlap with one another on host victim	16. guess_passwd	Attempt to steal user's information internally or externally
7. buffer overflow	overrides data interrupts data values due to insufficient bounds checking	17. Imap	Exploiting the vulnerability over webserver and mail server applications
8. load module	interruptions made by authorized persons on the local machine	18. Multihop	Attack happens when users try to access remote source.
9. perl	vulnerability exists based on root access on the local machine	19. phf	This prompt the remote-to-user attack against web servers
10. root-kit	exploiting automated vulnerability to attain direct attack on the system	20. spy	This is notified as a stealthy collection of user credentials through surfing and key loggers.

The majority of technical analysis [23] works with two prominent data-driven required for Internet-of-Things. They are known to be KDDCUP99 data and DARPA99 data. KDDCUP99 data has proven its quality of being efficient in determining the accuracy of malicious traces when compared to another database. The acronym of KDDCUP is Knowledge Discovery in data-sets. The KDDCUP99 data is empowered with diversified Intrusion Detection system-driven techniques. It has great deal of attention with TCP dump data to monitor network traffic. The data is mapped by approximation of five million records with forty-two attributes. The characteristics of these attributes is to determine whether network traffic is 'normal' or 'malicious.' The entire data had been segregated by three varied features and they are known to be 'Basic Features,' 'Traffic Features' and 'Content Features.'

The 'Basic Features' are familiar with computing the TCP/IP connection. To examine the time interval between connections, 'Traffic Features' are used. To validate the network traffic being neither 'normal' nor 'malicious' one, 'Content Features' are used. The entire data is used to examine the four different types of intrusions. They are known to be Denial-of- Service attacks, Probing attacks, User-to-Root attacks and Root-to-Local attacks. These attacks are sub-categorized with twenty-two attack types.

Table 3. Narrative of attribute sequence for 'basic features'

ATTRIBUTE SEQUENCE	ITS TYPE	STATE OF INTERACTION
duration	continuous	length of the connection
protocol_type	discrete	type of protocol ex: tcp, udp etc.
service	discrete	network service on the destination ex: telnet, http etc.
src_bytes	continuous	number of data bytes from source to destination
dst_bytes	continuous	number of bytes from destination to source
flag	discrete	normal or error status of the connection
land	discrete	1 is from/to the same host otherwise the connection is '0'
wrong_fragment	continuous	number of wrong fragments
urgent	continuous	number of urgent packets

The motivation behind this data [24] is to circumvent the designing in eminent network Intrusion Detection systems. The support of this data makes the researchers to train the varied algorithms with intent to improve their capability towards the accurate extraction of 'normal' and 'malicious' patterns. It provides better accuracy rate when compared to other data sets. Here, entire data represents binary format i.e., 0, 1. The binary '0' indicate the network traffic pattern as 'normal' and binary '1' indicate the network traffic pattern as 'malicious.' Each attack type sub-categorize with varied attack names which has its own characteristics. Table 2 represents the state of interaction for the attack names.

Table 4. Narrative of attribute sequence for 'content features' and its type

ATTRIBUTE SEQUENCE	ITS TYPE	STATE OF INTERACTION
hot	continuous	number of 'hot' indicators
num_failed_logins	continuous	number of failed login attempts
logged_in	discrete	1 if successfully logged in; '0' otherwise
num_compromised	continuous	number of 'compromised' conditions
root_shell	discrete	1 if root shell is obtained; otherwise '0'
su_attempted	discrete	1 if 'su root' command attempted; otherwise '0'
num_root	continuous	number of 'root' accesses
num_file_creations	continuous	Number of file creation operations
num_shells	continuous	Number of shell prompts
num_access_files	continuous	Number of operations on access control files
num_outbound_cmds	continuous	Number of outbound commands in an ftp session
is_hot_login	discrete	1 if the login belongs to "hot" list otherwise '0'
is_guest_login	discrete	1 is the login belongs to 'guest' otherwise login '0'

The sequence of attributes are categorized with three varied sets and they are known to be 'basic features,' 'content features' and 'traffic features.' The entire attribute is pre-empted by neither discrete nor continuous. Each attack connection is posed by forty-one attributes sequence. Additionally, one class label is determined to note neither 'normal' nor 'attack' exists. Each feature has its own characteristics [25]. The 'basic features' are used to extract all features from TCP/IP connection. This feature states with the implied delay in sequential analysis. The 'traffic features' were used to calculate time interval between 'same hosts' and 'same service' features. Here, the traffic pattern will be analyzed based on the time interval which exists that goes beyond two seconds. For example, is it is one minute, then 'traffic pattern' is notified as normal. In case of traffic pattern existing between two seconds, then it is notified as 'attack' pattern. Validation is done by recalculating the 'same host' and 'same service' preferably to a time window of two seconds. The 'content features' compute suspicious behavior in traffic pattern. Table 3, 4 and 5 determine the state of interaction for attribute sequence.

Table 5. Narrative of attribute sequence for 'traffic features' and its type

ATTRIBUTE SEQUENCE	ITS TYPE	STATE OF INTERACTION
count	continuous	number of connections to the same host as the current connection in the past two seconds
serror_rate	continuous	number of connections that have 'SYN' errors
rerror_rate	continuous	number of connections that have 'REJ' errors
same_srv_rate	continuous	number of connections to the same service
diff_srv_rate	continuous	number of connections to different services
srv_count	continuous	number of connections to the same service as the current connections in the past two seconds
srv_serror_rate	continuous	number of connections that have 'SYN' errors
srv_rerror_rate	continuous	number of connections that have 'REJ' errors
srv_diff_host_rate	continuous	number of connections to different hosts

The data which is widely adopted in evaluating the Intrusion Detection system is DARPA99 data. The function of this data is groomed by Secure Shell Handshake (SSH) from the cluster of flow-based Intrusion Detection Evaluation (IDEVAL) data. The characteristics of this data consist of an approximation of five million records with twenty-two attributes. The facet of SSH protocol is notified as an open-source protocol that prompts the system to access in a remote way. Moreover, Secure Shell Handshake protocol emphasizes for file tunnelling and transfers. DARPA99 data is desperately steady with SSH protocol. For understanding purpose, Table 6 represents the state of interaction for its attribute sequence.

Table 6. DARPA99 data represents the state of interaction for its attribute sequence

ATTRIBUTES	SEQUENCE OF ATTRIBUTES	STATE OF INTERACTION
Packet Length (forward)	1. min_fpktl 2. mean_fpktl 3. max_fpktl 4. std_fpktl	minimum forward mean forward maximum forward std dev /* standard deviation*/
Packet Length (backward)	5. min_bpktl 6. mean_bpktl 7. max_bpktl 8. std_bpktl	minimum backward mean backward maximum backward std dev /* standard deviation*/
Inter arrival time (forward)	9. min_fiat 10. mean_fiat 11. max_fiat 12. std_fiat	minimum forward mean forward maximum forward std dev /* standard deviation*/
Inter arrival time (backward)	13. min_biat 14. mean_biat 15. max_biat 16. std_biat	minimum backward mean backward maximum backward std dev /* standard deviation*/
duration		duration of the flow
proto		protocol (TCP, UDP)
total_fpackets		total forward packets
total_fvolume		total forward volume
total_bpackets		total backward packets
class		to identify attack or normal

The optimized classifier-based techniques project its compatibility to both signature-based detection techniques and Anomaly-based detection techniques. Some of the prominent techniques [26] prove its enhancement in fine-tuning the detection techniques with intent to achieve the basic requirements and they are known to be Clustering algorithm, Association rule, Genetic algorithm, Fuzzy logic, Classification algorithm, etc. The characterization of classification and clustering approach is found to be same in categorizing the facts into one or more classes based on features. The classification-based techniques are widely used in supervised learning techniques and their need to label the data for training and testing the data. Whereas, clustering-based techniques are widely used in unsupervised learning technique and here there is no need of training and testing the data.

The Synthetic Minority Oversampling technique (SMOTE) endorsed to establish the distinguished classification accuracy, based on the performance [27] of six varied classifiers analysis. The classifiers stage through the F1-scores by cross validating each classifier analysis. Here, three varied datasets are used and they are known to be KDD-99, NSL-KDD and UNSW-NB15. The classifiers used are Naïve Bayes, Support Vector Machine (SVM), Decision Tree, Random Forest, Neural Networks and K-means. The weighted F1-score achieved its overall accuracy from KDD-99 in an approximation of 94%. The overall weighted F1-score achieved its accuracy with the approximation of 83% and 86.5% for NSL-KDD and UNSW-NB15 data respectively.

The flow-based intrusion detection system has proposed [28] using neural networks. The subjective is to classify and detect the malicious network traffic patterns. The experimentation was executed with the support of DARPA99 data-set from MATLAB background. The experimentation has been split into two sets. In the first set, from the entire collection of DARPA data, only eight hundred samples were taken into account. The detection rate obtained at 92.7%, 94.2% and 91.1% in terms of trial-and-error basis and false positives are 3.6%, 3.4% and 5.1% respectively. In the second set, approximately twenty-one thousand samples were taken into account. The detection rate obtained is 95.4% and 99.4% and false positives are 4.6% and 0.58% respectively.

Some of the divergent approaches are developed with intent to boost up the classifier performance to achieve high detection accuracy. The structure with Support Vector Machine and Rough Set theory was proposed as hybrid approach [29] with the back-end of KDDCUP99 data. This method was effective to avoid redundancy over the features. The experimentation was split into two levels. In the first level, entire forty-one features were taken into account and its detection rate was achieved at 86.79% and false positive rate at 29.97%. In the second level, structure works with twenty-nine features in which the detection rate was 89.13% and its false positive rate was at 13.27% respectively. Based on the same structure, another approach had been experimented by reducing forty-one features to fifteen features [30]. The detection rate obtained was at 97.74% and its false positive rate was at 2.318% when compared to one hand approach of Support Vector Machine achieved at 97.73% and false positive rate at 2.311% respectively.

The Genetic algorithm-based detection approach was laid on eminent platform in analyzing the 'normal' and 'malicious' network traffic patterns. Some of the structures are discussed here. The structure of parallel genetic local search algorithm [31] was experimented based on the fuzzy rules. The detection rate obtained was at 96.3% and false alarm rate was at 0.29% simultaneously. The rule-based genetic algorithm was proposed [32] to categorize each attack class and its detection rate obtained at 83.5%. Another performance was countersigned by genetic algorithm through network sniffing [33] for accurate classification of network instances. The overall detection rate was attained at 99% and false positive rate at 3%.

Deep learning model is another structure that has the potential [34] to boost up the performance of cyber traffic in practical source. The common fact of learning model is to deliberate the accurate classification of 'normal' and 'malicious' instances. Here, some of the deep learning models are compared and the performance analysis made on NSL-KDD data-set to detect anomalies presence. The models are Inception Convolutional Neural Network (CNN), Deep Belief Network (DBN), and Bidirectional Long Short-Term Memory (BLSTM). The detection rate attained for each model was at 88.03%, 71.91% and 84.03%. The false positive rate for Inception model was at 28.576% normal traffic and its attack traces (DOS, Probe, R2L, U2R) incurred a range of 28.62%, 6.06%, 2.178%, 0.063% respectively. The false positive rate for BLSTM at 65.448% normal traffic and its attack traces (DOS, Probe, R2L, U2R) incurred at a range of 24.659%, 10.323%, 7.810%, 0.082%. The false positive rate for Deep Belief Network model at 46.533% normal traffic and its attack traces (DOS, Probe, R2L, U2R) incurred at a range of 25.2108%, 14.073%, 6.073% and 1.065% respectively.

Support Vector Machine (SVM) is one of widely used supervised machine learning classifier [35] technique, which is derived from the principle of structured risk minimization. The reason is, it creates the maximum hyperspace between two quantities and makes the researchers utilize this classifier technique neither in single-hands nor with hybrid approach with an intent to boost up the performance of classification techniques to attain better requirement. The experimentation was conducted using five different classification algorithms known to be Support Vector Machine, Artificial Neural Networks, K-means Nearest Neighbour, Naïve Bayes and Decision Tree. The detection rate was attained at a rate of 97.3285%, 95.7594%, 99.4403%, 89.5919%, 99.5594% and false alarm rate was attained at 2.6715%, 4.2406%, 0.5597%, 10.4081% and 0.4406% respectively.

The Efficient Data-Adapted Decision Tree algorithm was proposed to tune the SVM parameters by applying the Radial Basis Function with intent to boost up the performance of hybrid approach to attain the high detection rate by minimizing its false positives. The experimentation was conducted to attain its detection rate at 98.12% and false positives at a rate of 0.18%. The back propagation-based detection techniques proved its technical intelligence in categorizing the accurate classification of network traffic patterns. Moreover, its uniqueness helps in achieving the trustworthiness to overcome [37] presence in existence. The simulation was performed using MATLAB with the back-end of KDDCUP99 data-set. The cross-validation done through varied level of testing and its detection rate was attained at 95.6%, and false alarm rate at 4.4% in level one. In the second level of testing, the detection rate was attained at 73.9%, and false alarm rate at 26.1%.

The experimentation was conducted from the comparative analysis of four classifier techniques [38] using two varied data-sets. The Next Generation Intrusion Detection System data-set and KDDCUP99 data were used for experimentation. The four varied machine learning algorithms are SVM, Random Forest, J48, REP tree. The detection rate obtained using

KDDCUP99 data was at a rate of 95.93%, 99.57%, 99.90%, and 99.90%. The false positives were obtained at a rate of 0.001%, 0.0007%, 0.0003%, and 0.0002%. The detection rate obtained using Next Generation Intrusion Detection System data set was at 76.57%, 81.23%, 79.61% and 73.54%. The false positive obtained was at a rate of 17.59%, 14.84%, 20.12% and 19.42% respectively.

The supervised machine learning technique, which is most widely used for designing the Intrusion Detection systems known to be Extreme Learning Machine (ELM). The pattern of Extreme Learning Machine prompted as a single-layer feed forward network with an intent map that features using Constant Probability Distribution (CPD) function [39]. The experimentation was conducted to boost up the performance of Online Sequential Extreme Learning Machine (OS-ELM) using gain ratio evaluation method and the resulting analysis will be compared with other classifier algorithms. The lists of algorithms used for experimentation are Extreme Learning Machine (ELM), OS-ELM, Back Propagation Neural Network (BPNN) and Radial Basis Function (RBF). The detection rates were attained at 95.369%, 97.239%, 95.64%, and 96.034%. The false positives were attained at 6.067%, 5.897%, 4.374%, and 8.144% respectively.

The Levenberg-Marquard algorithm was proposed [40] towards Kernelized Extreme Learning Machine for designing the Intrusion Detection System. The simulation was conducted using KDDCUP99 data-set and its detection rate was obtained at 97.89%. The false alarm rate obtained at was 1.06% respectively. Based on the concept of OS-ELM, network traffic profiling was proposed [41] for designing the Intrusion Detection System. The motive is to reduce the time complexity and feature size. Experimentation was conducted using two varied data-sets and they are known to be Kyoto University benchmark data-set at 96.73% and its false alarm rate at 5.76%. On the other hand, the detection rate obtained by KDDCUP99 data-set is at 98.66% and its false alarm rate at 1.74%. Table 7 represents the methodologies involved in designing efficient Intrusion Detection Systems from the year 2008 to 2018.

Table 7. Methodologies involved in Intrusion Detection Systems

AUTHOR & YEAR	METHODOLOGY	DATASET	DETECTION RATE	FALSE ALARM RATE
Sagale et.al. [2014]	SVM-Naïve Bayes	KDDCUP99	99%	6.8%
Yuan et.al. [2010]	Radial Basis Function-SVM	KDDCUP99	93.7%	0.11%
Zaman et.al. [2009]	Fuzzy based Enhanced Support Vector Decision	KDDCUP99	99.66%	0.24%
Rathore et.al. [2016]	Support Vector Machine Random Forest J48 REP tree	KDDCUP99	95.93% 99.57% 99.90% 99.90%	0.001% 0.0007% 0.0003% 0.0002%
	Support Vector Machine Random Forest J48 REP tree	Next Generation IDS	76.57% 81.23% 79.61% 73.54%	17.59% 14.84% 20.12% 19.42%
Kausar et.al. [2012]	Principle Component Analysis- Support Vector Machine	KDDCUP99	99.46%	0.52%
Barapatra et. al. [2008]	Training MLP Neural Network	KDDCUP99	81.96%	8.51%
Sadiq Ali Khan [2011]	Rule based IDS- Genetic Algorithm	KDDCUP99	94.19%	2.75%
Abuadullah et. al [2014]	Flow based Anomaly Intrusion Detection System	DARPA dataset with 22,000 samples	Experiment 1 : 92.7% 94.2% 91.1% Experiment 2: 95.4% 99.4%	3.6% 3.4% 5.1% 4.6% 0.58%
Kasun et. al [2013]	Extreme Learning Machine- Auto Encoder	MNIST 60,000 samples for training 10,000 samples for testing	99.03%	n.a
Muda et. al [2011]	One-R & K-means clustering	KDDCUP99	99.26% & 99.33%	2.73%
Yuangcheng et. al [2018]	1. Online Sequential Extreme Learning Machine	Advanced Metering Infrastructure	97.239%	5.89%
	2. Extreme Learning Machine		95.369%	6.06%
	3. Back Propagation Neural Network		95.64%	4.57%
	4. Radial Basis Function		96.03%	8.14%
Praneet et. al [2016]	Efficient Proactive artificial Immune System	Real valued representation	Approximation of 83.72% -100%	0.68%-0.81%
Sadasivan et.al [2017]	Adaptive Rule based Multiagent IDS	KDDCUP99	97.4%	2.24%

2.4. EVOLUTIONARY-BASED APPROACHES

The quality of evolutionary-based detection approaches has proven their characterization in designing the eminent Intrusion Detection Systems [42]. These types of detection approaches are far better than existence. The behaviors of these approaches were designed-based on real characteristics of mammals, species, etc. Some of the fame approaches are Ant Colony Optimization (ACO), Particle Swarm Optimization (PSO), Grey Wolf Optimization (GWO), Bees Colony Optimization (BCO), Butterfly Optimization Algorithm (BOA), etc. The objective of these detection approaches is to provide the high detection accuracy by minimizing its false leads. The extraction of 'normal' and 'malicious' traffic pattern is not a trouble-free quest. Some of the approaches achieved its ideal in accurate classification of instances to a certain extent.

The combined approach of classifier and evolutionary models proved their sufficient optimum by reducing the false alarms effectively. This leads to achieve a good accuracy rate in classification of instances. Particle Swarm Optimization (PSO) technique is the simplest one, which were used to optimize the performance of Multiple Criteria Linear Programming (PSO-MCLP) as a hybrid approach [43] with intent to attain the accuracy of attacks detection. KDDCUP99 data were used for experimentation. The simulation analysis was performed by comparing PSO-MCLP, MCLP, SVM and C5.0 classifier techniques. The detection rate was obtained at 99.13%, 97.46%, 98.07%, and 98.38%. The false alarm rate attained 0.019%, 0.036%, 0.027%, and 0.042%. Based on the pheromone behavior of ants moving towards the destination path, Ant Colony Optimization technique was framed.

The Ant Colony Optimization-based feature selection was framed [44] to design the efficient detection system. The performance of this approach attained its detection rate at 98.9% and false positive at 2.59%. The Clustering-based Self Organizing Ant Colony Networks (CSOACN) [45] was proposed using hybrid approach of SVM and Ant Colony networks. The objective of this existence is to design high dimensional data for designing the real-time Intrusion Detection Systems. The overall detection accuracy was obtained at 86.10% and its false positive was obtained at 0.36%.

The hybrid classifier model of Spider Monkey Optimization (SMO) and Deep Neural Networks (DNN) were developed for designing the Intrusion Detection System [46]. The experimentation was conducted using KDDCUP99 data and NSL-KDD data. The resulting analysis of this approach was compared with hybrid approach of Principle Component Analysis and Deep Neural Network (PCA+DNN) and also with one-hand approach of Deep Neural Network (DNN). The detection rate obtained using NSL-KDD was at 99.4% for SMO+DNN, 93.8% for PCA+DNN and 91.4% for DNN. The detection rate obtained using KDDCUP99 at 92.8% for SMO+DNN, 89.8% for PCA+DNN and 90.9% for DNN.

Another novel evolutionary-based detection approach was developed [47] based on the real behavior of western wolves. The major characteristic of Grey Wolf Optimization (GWO) algorithm is termed to be heterogeneity, i.e., made of a diversified set of elements to the higher-level instructional practices for experimentation purpose. Moreover, the huge elements were gathered together, which appear to be live in the pack. Using the wild candid nature of Grey wolves, algorithm was designed as resilience nature intent to execute a stealthy operation towards prey. Finally, it shortens the distance to reach the destination. The simulation analysis of GWO is far better than homogeneous families of optimization techniques (ACO, PSO etc).

The Improved Grey Wolf Optimization with Support Vector Machine was developed as hybrid technique (IGWO-SVM) [48] for feature extraction from the real-time data events. The collection of students' data and their subjects allotted were used for experimentation. The classification accuracy is to update current position of students, which appeared at a rate of 87.36% and shifted its irrelevant features, i.e., false leads at 5.26% respectively. A further approach of Improved Grey Wolf Optimization technique with Kernelized Extreme Learning Machine (IGWO-KELM) was designed for medical diagnosis [49]. The goal of the experimentation was to update the optimal feature extraction for better classification. The simulation analysis was performed using two varied data-sets and they are known to be Wisconsin Diagnostic Breast Cancer (WDBC) and Parkinson data set. The classification accuracy attained using WDBC data at 95.78% and accuracy of Parkinson data were at 97.45%.

Based on the meritorious utilization of Support Vector Machine technique in designing the eminent Intrusion Detection System, hybrid approach of Support Vector Machine with Grey Wolf Optimizer (SVMGWO) [50] were experimented. The objective of this approach was to improve the performance of Support Vector Machine with the additional support of Grey Wolf Optimization technique to produce a high detection rate by reducing its false leads to a minimal level. The simulation analysis will be compared with other existence. The approximation of twenty-five thousand instances from KDDCUP99 data were taken into account for analysis of class-wise attributes. The detection accuracy was attained at 99.98% and its false alarm rate was attained at 0.01% compared with Efficient Data-Adapted Decision Tree algorithm [EDADT] which attained an accuracy of 98.12% and its false alarm rate at 0.18% respectively. This shows that SVMGWO had proven its eminence in detection of accurate attack instances by reducing its false leads.

Another approach, SVMGWO showcases its efficiency by achieving the accurate classification of instances from the overall population. The performance evaluation [51] will be compared with LibSVM and SMO algorithms. The characteristic of LibSVM and SMO are different, but both are used to boost up the performance of SVM. Sequential Minimal Optimization (SMO) is used to improve the support vector classifier and LibSVM, which is the wrapper class of SVM. The detection accuracy was attained by SVMGWO with an approximation of 99.8902% - 99.9109% and its false alarm rate was at 0.1098%-0.0899%. The detection accuracy attained by LibSVM was with an approximation of 99.7093%-99.788% and its false alarm rate at 0.2907%-0.212%. Table 8 represents methodologies of evolutionary and hybrid approaches (classifier-evolutionary based) for designing the Intrusion Detection Systems.

Table 8. Methodologies of evolutionary and its hybrid detection approaches

AUTHOR & YEAR	METHODOLOGY	DATASET	DETECTION RATE	FALSE ALARM RATE
Nilesh et.al [2020]	Particle Swarm Optimization	NSL-KDD	99.26%-99.32%	n/a
Chung et.al [2020]	Simplified Swarm Optimization	KDDCUP99	93.3%	n/a
Alazzam et.al [2020]	Pigeon Inspired Optimizer	(i) KDDCUP99 (ii) NSL-KDD (iii) UNSW-NB15	97.4% 86.9% 91.3%	0.097-0.001% 0.064-0.0008% 0.052-0.0004%
Sweta et.al [2020]	Hybrid PCA – Firefly Optimization 1. KNN-PCA & Firefly 2. NB -PCA & Firefly 3. RF-PCA & Firefly 4. SVM-PCA & Firefly 5. XGBoost-PCA & Firefly	Kaggle	99.4% 84.2% 99.8% 97.5% 99.9%	n/a
Zhou et. al [2019]	Bat Algorithm and Correlation based Feature Selection	CIC-IDS2017	94.04%	2.38%
Mohammadi et. al [2019]	Cuttlefish	KDDCUP99	95.23%	1.65%
Acharya et.al [2018]	Intelligent Water Drops Algorithm using SVM	KDDCUP99	91.35%	3.35%
Keshtgary et.al [2018]	Feature Selection & Clustering using SVM- K Medoids- Naïve Bayes	KDDCUP99	90.1%	6.36%
Khammassi et.al [2017]	Genetic Algorithm-Decision Tree	(i) UNSW NB-15 (ii) UGR dataset	99% 98.8%	5.2% 0.01%
Vidhya et.al [2017]	Hybrid SVM with Grey Wolf Optimizer	KDDCUP99	99.98%	0.01%
Aslahi et. al [2016]	Genetic Algorithm	KDDCUP99	97.03%	1.7%
Aghdam et.al [2016]	Ant Colony Optimization	KDDCUP99	98.9%	2.59%
Enache et.al [2015]	Improved Bat Algorithm with Support Vector Machines	NSL-KDD	95.05%	4.4%
Ghanam et.al [2015]	Multi-Start meta heuristics method & Genetic Algorithm	NSL-KDD	96.1%	0.033%
Atefi et.al [2013]	Support Vector- Genetic Algorithm	KDDCUP99	99.49%	1.78%
Li et.al [2012]	Clustering & Ant Colony Optimization &SVM	Real-World	99.62%	n/a

The feature selection using Pigeon Inspired Optimizer was proposed [52] by adjusting the velocity and position to travel towards the prey. The experimentation was performed by deploying three variant data sets and they are known to be KDDCUP99, NSL-KDD and UNSW-NB15. The detection rate using KDDCUP99 was attained at a rate of 97.4% and false positive at an approximation of 0.097%-0.001%. The detection rate using NSL-KDD was attained at a rate of 86.9% and false positive at approximation of 0.064%-0.0008%. The detection rate using UNSW-NB15 was attained at a rate of 91.3% and false positive at an approximation of 0.052%-0.0004% respectively.

CONCLUSION

This chapter acknowledges the existence of anomaly detection approach to overwhelm the presence of intrusions. Moreover, this chapter also elaborately presented the details based on classifier-based, evolutionary-based and also its hybrid approaches. It also discusses about how detection techniques got trained for accurate classification of instances by reducing its false leads at minimum extent. This chapter paved the way to a state-of-the-art knowhow about enhanced optimization techniques to withstand over the cyber research challenge.

REFERENCES

[1] Ahmad Karim, Rosli Bin Salleh, Muhammad Shiraz, Syed Adeel Alishah, Irfan Awan and Nor Badrul Anuar, Botnet Detection Techniques: Review, Future Trends and Issues (*Journal of Zhajiang University SCIENCE C,* 2014), 15(11), 943-983.

[2] Doron Jeffrey Samuel, Intrusion Detection System-Features, Classification of Attacks and Difficulties (*Imperial Journal of Interdisciplinary Research*, 2016), 2(4), 689-692.

[3] Pedro Garcia-Teodoro, Jesus E. Diaz-Verdejo, Gabriel Macia-Fernanadez, Enrique Vazquez, Anomaly-based Network Intrusion Detection: Techniques, Systems and Challenges (*Computers and Security*, 2009), 28(1-2), 18-28.

[4] Amria Sayed Abdel-Aziz, Aboul Ella Hassanien, Ahmad Taher Azar, Sanaa El-Ola Hanafi, Machine Learning Techniques for Anomalies Detection and Classification (*Advances in Security of Information and Communication Networks*, 2013), 381, 219-229.

[5] Ansam Khraisat, Iqbal Gondal, Peter Vamplew and Joarder Kamruzzaman, Survey of Intrusion Detection Systems: Techniques, Datasets and Challenges (*Cybersecurity*, Springer, 2019), 2(20), 2-22.

[6] Jyothsna Veeramreddy, Rama Prasad V.V, Koneti Munivara Prasad, A Review of Anomaly based Intrusion Detection Systems (*International Journal of Computer Applications*, 2011), 28(7), 26-35.

[7] Asieh Mokarian, Ahmad Faraahi, Arash Ghorbannia Delavar, False Positives Reduction Techniques in Intrusion Detection Systems - A Review (*International Journal of Computer Science and Network Security*, 2013), 13(10), 128-134.

[8] Salman M, Badiardjo B, Kalamullah Ramli, Key Issues and Challenges of Intrusion Detection and Prevention System: Developing Proactive Protection in Wireless Network Environment (*International Journal of Electrical, Computer, Energetic, Electronic and Communication Engineering*, 2011), 5(5), 659-662.

[9] Halim Mohamad tahir, Wael Hasan, Abas Md Said, Nur Haryani Zakaria, Norliza katuk, Nur Farzana Kabir, Mohd Hasbullah Omar, Osman Ghazali, Noor Izzah Yahya, Hybrid Machine Learning Technique for Intrusion Detection System (*Proceedings of the 5th International Conference on Computing and Informatics*, 2015), 209, 464-472.

[10] Zenghui Liu, Yingxu Lai, A Data Mining Framework for Building Intrusion Detection Models based on IPV6 (*ISA'09: Proceedings of*

the 3rd *International Conference and Workshops on Advances in Information Security and Assurance*, 2009), 5576, 608-618.
[11] Kavita Patond, Pranjali Desmukh, Survey on Data Mining Techniques for Intrusion Detection System (*International Journal of Research Studies in Science, Engineering and Technology*, 2014), 1, 93-97.
[12] Constantinos Kolias, Georgios Kambourakis, Manoli Maragoudakis, Swarm Intelligence in Intrusion Detection: A Survey (*Computers and Security*, 2011), 30(8), 625-642.
[13] Amudha Arul, Karthik Subburathinam, Sivakumari S, A Hybrid Swarm Intelligence Algorithm for Intrusion Detection using significant features (*The Scientific World Journal*, 2015), 2015, 1-15.
[14] Mohamed Faisal Elrawy, Ali Ismail Awad, Hesham F.A. Hamed, Intrusion Detection System for IOT-based Smart Environments: A Survey (*Journal of Cloud Computing*, 2018), 7(21), 1-20.
[15] Shaker El-Sappagh, Ahmed Saad Mohammed, Tarek Ahmed Alshehtawy, Classification Procedures for Intrusion Detection Based on KDDCUP99 Dataset (*International Journal of Network Security & its Applications (IJNSA)*, 2019), 11(3), 21-29.
[16] Preeti Aggarwal, Sudhir Kumar Sharma, An Empirical Comparison of Classifiers to Analyze Intrusion Detection (*fifth International Conference on Advanced Computing & Communication Technologies* (ACCT), 2015).
[17] Audrey Gendreau, Michael Moorman, Study of Intrusion Detection Systems towards an End to End Secure Internet of Things (IEEE 4th International Conference on Future Internet of Things and Cloud, *IEEE Computer Society*, 2016), 84-90.
[18] Guofei Gu, Phillip Porras, Vinod Yegenswaran, Martin Fong, Wenke Lee, Bot Hunter: Detecting Malware Infection through IDS-Driven Dialog Correlation (In: *Proceedings of 16th USENIX Security Symposium, USENIX Association,* Berkeley, CA:USA, 2007),12, 167-182.
[19] Guofei Gu, Roberto Perdisci, Junjie Zhang, Wenke Lee, Bot Miner: Clustering Analysis of Network Traffic for Protocol and Structure

Independent Botnet Detection (*SS'08: Proceedings of the 17th Conference on Security Symposium*, 2008), 139-154.

[20] Mitchell D Silva, Deepali R Vora, Comparative Study of Data Mining Techniques to Enhance Intrusion Detection (*International Journal of Engineering Research and Applications*, 2013), 3, 1267-1275.

[21] SeRgio S.C.Silva, Rodrigo M.P.Silva, Raquel C.G. Pinto, Ronaldo M. Salles, Botnets: A Survey (*Computer Networks: The International Journal of Computer and Telecommunications Networking*, 2013), 57(2), 378-403.

[22] Robert Mitchell, Ing-Ray Chen, Behavior Rule Specification-based Intrusion Detection for Safety Critical Medical Cyber Physical Systems (*IEEE Transactions on Dependable and Secure Computing*, 2015), 12(1), 16-30.

[23] Suchet Sapre, Pouyan Ahmadi, Khondkar Islam, A Robust Comparison of the KDDCUP99 and NSL-KDD IoT Network Intrusion Detection Datasets through Various Machine Learning Algorithms, Aspiring Scientists Summer Internship Program (*Mason Journals*, the College of Science, George Mason University, arXiv:1912.13204V1[CS.LG], 2019).

[24] Mahbod Tavallaee, Ebrahim Bagheri, Wei Lu, and Ali A.Ghorbani, A Detailed Analysis of the KDDCUP99 dataset (*Proceedings of the IEEE Symposium on Computational Intelligence in Security and Defense Applications* (CISDA 2009), 2009), 53-58.

[25] Safaa O. Al-mamory, Firas S. Jassim, Evaluation of Different Data Mining Algorithms with KDDCUP99 Dataset (*Pure and Applied Sciences, Journal of Babylon University*, 2013), 21(8), 2663-2681.

[26] Jaiganesh V, Mangayarkarasi S and Sumathi P, Intrusion Detection Systems: A Survey and Analysis of Classification Techniques (*International Journal of Advanced Research in Computer and Communication Engineering*, 2013), 2, 1629-1635.

[27] Abhishek Divekar, Meet Parekh, Vaibhav Savla, Rudra Mishra, Mahesh Shirole, Benchmarking datasets for Anomaly-based Network Intrusion Detection : KDDCUP99 alternatives (*IEEE International*

Conference on Computing, Communication and Security (ICCCS-2018), 2018).

[28] Yousef Abuadlla, Goran Kvascev, Slavko Gajin, Zoran Jovanovic, Flow-based Anomaly Intrusion Detection system using Two Neural Network stages (*Computer Science and Information Systems*, 2014), 11(2), 601-622.

[29] Rung-Ching Chen, Kai-Fan Cheng, Ying-Hao Chen and Chia-Fen Hsieh, Using RoughSet and Support Vector Machine for Network Intrusion Detection System (First Asian Conference on Intelligent Information and Database Systems, *IEEE Computer Society*, 2009), 465-470.

[30] Ravinder Reddy R, Kavya B, Ramadevi Y, A Survey on SVM classifiers for Intrusion Detection (*International Journal of Computer Applications*, 2014), 98(14), 38-44.

[31] Mohammad Saniee Abadeh, Jafar Habibi, Zeynab Barzegar, Muna Sergi, A Parallel Genetic Local Search algorithm for Intrusion Detection in Computer Networks (*Engineering Applications of Artificial Intelligence*, 2007), 20(8), 1058-1069.

[32] Shaik Akbar, Dr. Nageswara Rao k, Dr. Chandlal J A, Implementing Rule based Genetic Algorithm as a Solution for Intrusion Detection System (*International Journal of Computer Science and Network Security*, 2011), 11(8), 138-144.

[33] Salah Eddine Benaicha, Saoudi Lalia, Salah Eddine Bouhouita Guermecha, Quarda Lounis, (*Science and Information Conference (SAI), IEEE*, 2014), 564-568.

[34] Safaa Laqtib, Khalid El Yassini, Moulay Lahren Hasnaoui, A Technical Review And Comparative Analysis of Machine Learning Techniques for Intrusion Detection systems in MANET (*International Journal of Electrical and Computer Engineering*, 2020), 10(3), 2701-2709.

[35] Ajayi Adebowale, Idowu S.A, Anyachie Amarachi, Comparative Study of Selected Data Mining Algorithms used for Intrusion Detection (*International Journal of Soft Computing and Engineering*, 2013), 3(3), 237-241.

[36] Nadiammai G.V, Hemalatha M, Efficient approach toward Intrusion Detection System using Data Mining techniques (*Egyptian Informatics Journal*, 2014), 15(1), 37-50.

[37] Indraneel Mukhopadhyay, Mohuya Chakraborty, Chakrabarti S, Tanusree Chatterjee, Back Propagation neural network approach to Intrusion Detection System (*International Conference on Recent Trends in Information Systems*, 2011), 303-308.

[38] Kamran Siddique, Zahid Akhtar, Farrukh Aslam Khan, Yangwoo Kim, KDDCUP99 Datasets: A Perspective on the Role of Datasets in Network Intrusion Detection Research (*IEEE Computer Society*, 2019), 41-51.

[39] Yuangcheng Li, Rixuan Qiu, Sitong Jing, Intrusion Detection System using Online Sequence Extreme Learning Machine (OS-ELM) in (Advanced Metering Infrastructure of Smart Grid, *PLUS ONE*, 2018), 13(2), 1-16.

[40] Jaiganesh V, Sumathi P, Kernelized Extreme Learning Machine with Levenberg-Marquard Learning Approach towards Intrusion Detection (*International Journal of Computer Applications*, 2012), 54(14), 38-44.

[41] Raman Singh, Harish Kumar, Singla R.K, An Intrusion Detection System using network traffic profiling and Online Sequential Extreme Learning Machine (*Expert Systems with applications*, 2015), 42(22), 8609-8624.

[42] Weifeng Sun, Min Tang, Lijun Zhang, Zhiqiang Huo, Lei Shu, A Survey of using Swarm Intelligence Algorithms in IoT (*Special Issue Surveys of Sensor Networks and Sensor Systems Deployments, Sensors*, 2020), 20(5), 1420, 1-27.

[43] Seyed Mojtaba Hosseini Bamakan, Behnam Amiri, Mahboubeh Mirzabagheri, Yong Shi, A New Intrusion Detection Approach using PSO based Multiple Criteria Linear Programming (*Procedia Computer Science*, 2015), 55, 231-237.

[44] Mehdi Hosseizadeh Aghdam, Peyman Kabiri, Feature Selection for Intrusion Detection System using Ant Colony Optimization (*International Journal of Network Security*, 2016), 18(3), 420-432.

[45] Wenying Feng, Qinglei Zhang, Gongzhu Hu, Jimmy Xiangji Huang, Mining Network Data for Intrusion Detection through combining SVMs with Ant Colony Networks (*Future Generation Computer Systems*, 2014), 37, 127-140.

[46] Neelu Khare, Preethi Devan, Chiranji Lal Chowdhary, Sweta Bhattacharya, Geeta Singh, Saurabh Singh and Byungun Yoon SMO-DNN: Spider Monkey Optimization-Deep Neural Network Hybrid Classifier model for Intrusion Detection (*Deep Neural Network and their Applications*, 2020), 9(4), 692, 1-18.

[47] Seyedali Mirjalili, Seyed Mohammad Mirjalili, Andrew Lewis, Grey Wolf Optimizer (*Advances in Engineering Software*, 2014), 69, 46-61.

[48] Yan Wei, Ni Ni, Dayou Liu, Huiling Chen, Mingjing Wang, Qiang Li, Xiaojun Cui, Haipeng Ye, An Improved Grey Wolf Optimization Strategy Enhanced SVM and its application in Predicting the Second Major (*Mathematical Problems in Engineering*, Hindawi, 2017) 2017, 1-12.

[49] Qiang Li, Huiling Chen, Hui Huang, Xuehua Zhao, ZhenNao Cai, Changfei Tong, Wenbin Liu, Xin Tian An Enhanced Grey Wolf Optimization based Feature Selection Wrapped Kernel Extreme Learning Machine for Medical Diagnosis (*Computational and Mathematical methods in Medicine*, 2017), 2017, 1-15.

[50] Vidhya Sathish, Dr. Sheik Abdul Khader, Improved Detection Host based on Hybrid SVM using Grey Wolf Optimizer (*International Journal of Security and its Applications*, 2017), 11(9), 59-72.

[51] Vidhya Sathish, Dr. Sheik Abdul Khader, Enhanced Hybrid model of Support Vector-Grey Wolf Optimizer Technique to improve the Classifier's Detection Accuracy in designing the efficient Intrusion Detection model (*Asian Journal of Applied Sciences*, 2016), 4(1), 135-148.

[52] Hadeel Alazzam, Ahmad Sharieh, Khair Eddin Sabri, A Feature Selection Algorithm for Intrusion Detection System based on Pigeon Inspired Optimizer (*Expert Systems with Applications*, 2020), 148, 1-13.

In: Anomaly Detection
Editors: Saira Banu et al.
ISBN: 978-1-53619-264-3
© 2021 Nova Science Publishers, Inc.

Chapter 3

Anomaly Detection and Applications

Huichen Shu[*], *PhD*
Clausthal University of Technology, Germany

Abstract

Outlier Detection is a newborn, but one of the well-studied domains. The aim of this chapter is to give a heuristic introduction to outlier's detection in two parts. Firstly, the structure of traditional outlier detection algorithms is presented roughly. Besides, it reviews the adoption of each method to detect outliers with different tools, namely, Python or with RapidMiner. Assumptions of the typical traditional outlier detection algorithms are probabilistic and statistical, linear, proximity-based, and high dimensional methods. In this chapter, there is also a successful attempt at Integrated models with higher performance and generalization. Towards the end, the latest survey about Deep Anomaly Detection Techniques is also exhibited.

---

[*] Corresponding Author's E-mail: huichen.sophia@gmail.com.

3.1. INTRODUCTION OF OUTLIER DETECTION

Outliers are also called anomalies, abnormalities, or deviants during data mining and statistics literature. Usually, anomaly data points would be firstly deleted or replaced before prediction or classification in preprocessing. However, there are still potential values in outliers because of some machine malfunctions, or transportation failures. Some extreme unexpected features run obviously or wholly deviant from the regular samples. Similar to binary classifications, anomaly data points - as one class in anomaly detection - are sporadic than positive observations in binary classifications. In other words, when we do some binary classifications, whose positive sample volume is exceptionally sparse, anomaly detection methods are also practicable.

Recognizing these anomalies successfully and accurately would help a lot in many fields; it may raise the quality of production, reduce productive processing costs; monitor extreme weather changes, predict a hurricane or El Nino, diagnose a rare disease, assist in choosing a suitable treatment as early as possible, or even detect the illegal website spider automatically and in real-time, to work as a part of a firewall for websites.

Although the study of anomaly detection is still a young branch, there are already many well-developed techniques. In this paper, several statistical methods before the popularity of computers are described in section 1.2; remarkable linear models, such as regression algorithms and principal component analysis are in section 1.3; the foundation of the proximity-based method is locality, namely clusters, distance and density in section 1.4; high-dimensional methods are also briefly reviewed in section 1.5. As shown in each part, different methods are generated by a specific mechanism. They usually focus only on unit characters like local or global outliers. Nevertheless, the performance of the single simple models is usually poor so a complicated combination model ordinarily has higher applicability as shown in section 1.6. In the end, section 1.7, a summary of this chapter and future development are both included. Only two models from all applications in this chapter are acceptable, namely the local outlier factor model in section 1.4 and the integrated model in section 1.6.

Technically, there are many forms to implement, for instance, detecting with some software programs like Matlab, SPSS, or RapidMiner, or in R or Python. With an aim to improve efficiency and cross-platform capability, it is advocated to design a model with software like SPSS or RapidMiner and then model it with a Scripting language. Implementation in Python or R is nowadays rarely a topic. They two have respective advantages and disadvantages. For instance, when a bigger cache is needed, or when our program needs to cooperate with others like transmit pieces of information to another service, Python is much convenient. At the same time, R has its priority—strong visualization ability, and a complete R ecosystem like CRAN, Bioconductor, and Github. However, with the continuous improvement of the underlying communication technology of the compilation environment, R is also compliable in Python and many software programs like RapidMiner and SPSS. The main bottlenecks for R are small memory, and slow computing speed. Nowadays, there are many prepared tools to speed up R, like pqR, renjin, FastR, Riposte, etc.

Back in the nineteenth century, computation depended still mainly on manual calculation so the previous theories were established primarily on statistical basis. The likelihood of observations fits by probabilistic and statistical algorithms, such as detecting extreme univariate values with Tail inequalities and Statistical-Tail Confidence Test. Alternatively, extreme-value analysis is also active in multivariate problems. Depth-based, deviation-based, and angle-based methods are compelling in it.

3.2. APPLICATION WITH MUSK DATASET

Relationshionship of two attributes from the musk dataset is in Figure 1.

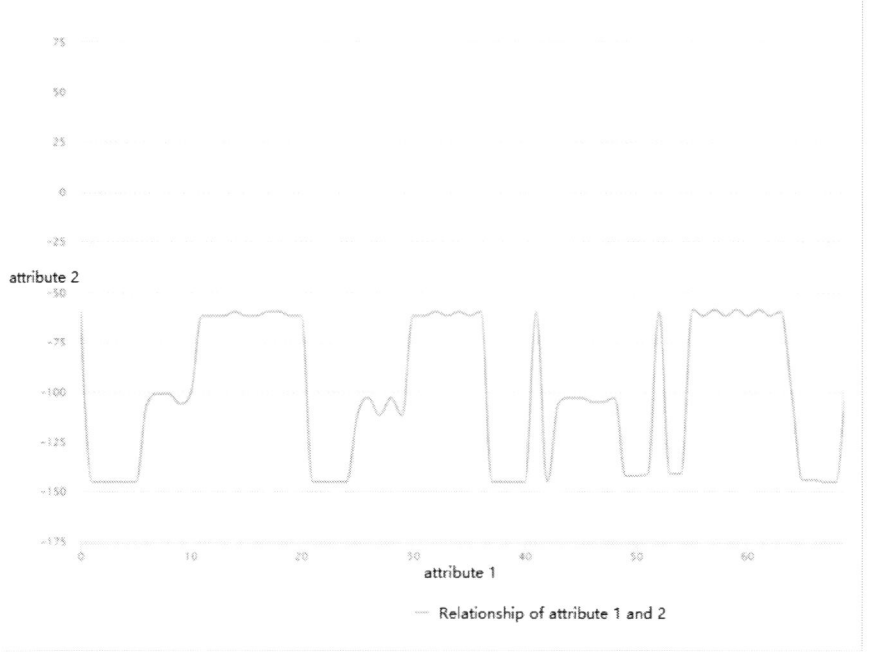

Figure 1. Attributes from MUSK dataset.

Let X be an arbitrary random variable. The expected value is the average of observations and the given attribute is reasonable in an interval between -60 and -130, namely the acceptable deviant is only up to 70. The anomaly score of X is the ratio of X variance and deviation squared according to Chebychev Inequality. X is an anomaly when its score is larger than the probability of |X-E(X)| greater than 70.

3.3. Linear Models Introduction

In realistic scenarios, there are usually great relationships among attributes. These models are ordinarily linear, referred to as regression. Outliers in linear models are the observations that deviate substantially from this regression model. However, the numerical evaluation of the deviation is defined by each model. And the definition affects the model's performance.

3.4. INTRODUCTION OF PRINCIPAL COMPONENT ANALYSIS (PCA) IN OUTLIER DETECTION

A typical application of the regression model is to detect outliers in time-series data and dependent variables from independent variables. A popular method is the principal component analysis. Eventual PCA in outlier detection [1] is a method to decrease features with the smallest loss of original information. Then the rest becomes the most critical dimension that can concisely and efficiently describe the original dataset.

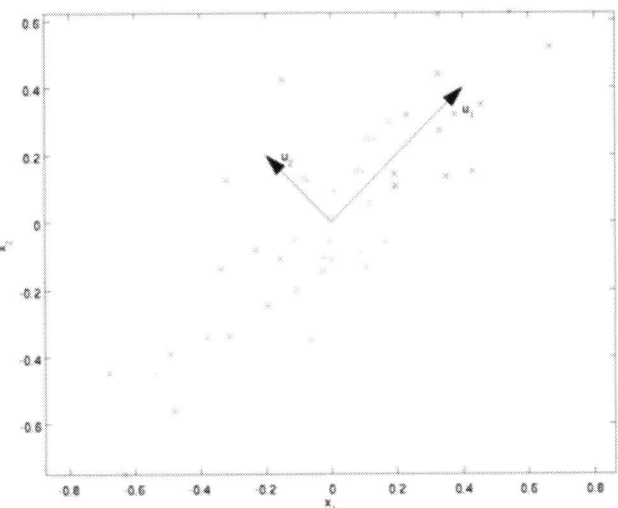

Figure 2. Application of pca method in outlier analysis.

As shown in Figure 2, the points' distributions are along two vectors namely, u1 and u2. Intuitively, there are more samples in u2 direction; the PCA model would regard u2 direction as an output to present the original dataset. The point of PCA is to find proper directions like u1 and u2 by the covariance matrix. Logically, points that are generated with the same principle would match the same directions. So the samples do not match the directions that most observations match and are hence regarded as Outliers.

Eventually, the PCA is already prepared for the decomposition of the sub package from Python's sklearn package. However, PCA in Outlier detection still needs to count the major components and the minor components of samples. After that, outliers are compared and distinguished. The samples, which have the most different components, are outliers.

3.5. APPLICATION WITH VOWELS DATASET

An example of the PCA model is compiled in Python with the original Japanese Vowels (Vowels) [2] dataset from UCI machine learning repository. This implementation uses the existinf = g PCA method in sub package "decomposition" of "sklearn," and then calculate major and minor components of observations like in Table 1.

This model takes no much time to end. In Table 2 demonstrates its result.

This model has high accuracy because PCA is an expert to find the majority. However, other evaluators like AUC and Recall are inferior. The reason may lie in the loss of critical potential features. Detailed features of Japanese vowels are unclear; upon this ambiguity, perhaps some crucial information to distinguish anomalies misses again. However, as an unsupervised dimensionality reduction method, the PCA model is simple to calculate and easy to implement because of the principal components orthogonal can eliminate the factor that affects the original data components.

Table 1. Importing Data Set

Importing dataset
from sklearn.preprocessing import StandardScaler from sklearn.decomposition import PCA scaler = StandardScaler() scaler.fit(vowels_features) vowels_scaledFreatures = scaler.transform(vowels_features) vowels_scaledFreatures = pd.DataFrame(vowels_scaledFreatures)
Training the model
pca = PCA() pca.fit(vowels_scaledFreatures) transformed_data = pca.transform(vowels_features) lambdas = pca.singular_values_ M = ((transformed_data*transformed_data)/lambdas) q = 5 print("Explained variance by first q terms: ",sum(pca.explained_variance_ratio_[:q])) q_values = list(pca.singular_values_ < 45) r = q_values.index(True) major_components = M[:,range(q)] minor_components = M[:,range(r, 12)] major_components = np.sum(major_components, axis=1) minor_components = np.sum(minor_components, axis=1) components = pd.DataFrame({'major_components': major_components,   'minor_components': minor_components}) c1 = components.quantile(0.99)['major_components'] c2 = components.quantile(0.99)['minor_components']
Testing the model
def classifier(major_components, minor_components): major = major_components> c1 minor = minor_components> c2 return np.logical_or(major,minor) data = scaler.transform(vowels_features) transformed_data_test = pca.transform(data) y_test = transformed_data lambdas_test = pca.singular_values_ M_test = ((y*y)/lambdas) major_components_test = M_test[:,range(q)] minor_components_test = M_test[:,range(r, 12)] major_components_test = np.sum(major_components_test, axis=1) minor_components_test = np.sum(minor_components_test, axis=1)
Evaluation

from sklearn.metrics import roc_auc_score
 	
results_train = classifier(major_components = major_components, minor_components =
    minor_components)
results_test = classifier(major_components = major_components_test, minor_components =
    minor_components_test)
#results_train == results_test

Table 2.

	true false	true true	class precision
pred. false	1391	41	97.14%
pred. true	15	9	37.50%
class recall	98.93%	18.00%	

3.6. INTRODUCTION OF ONE CLASS SUPPORT VECTOR MACHINES (SVM)

Support Vector Machines is a linear and logistic model to find a boundary to classify samples. One-Class SVM is an expansion of SVM for nonlinear outlier detection [3]. It means precisely a single class in original data. These points, which are separated by the One-Class SVM boundary are outliers.

Different from the PCA model, the SVMs model has no local minimum problem because there is only one class. The SVMs model uses gradient-descent to find the distribution of data. Not all zero gradients from original data are an extreme value, maximum or minimum, but second-derivative. With different kern functions, we can segregate observations with various boundaries. All boundaries may fit significantly to classify. However, sometimes the mechanism of outlier generation is still uncertain, which means the definition of a boundary is ambiguous. With mathematical application of different kern functions, the boundary may be detected.

One Class SVM is already well-packed in the "svm" subpackage of the "sklearn" package in Python. At the same time, this model is also beloved in software RapidMiner with an operator named "One-Class LIBSVM anomaly Score."

3.7. APPLICATION WITH RAPIDMINER

RapidMiner visualizes reading, selecting, filter, outputs, and various basic operations. We can draw an operation into the work area. Therefore, it is a convenient data analysis tool implemented in Java. So the Java environment is needed before installing RapidMiner. A unique toolkit in it for outlier detection is called "Anomaly Detection," which needs to be downloaded in the embedded market. At the same time, there are many other tools in the market such as plugins that support Python or R. Because it is graphical and supports other analytic languages, it is particularly convenient in application scenarios such as viewing data distribution and model design, for instance, the implementation of outlier detection for the Vowels dataset in Figure 3.

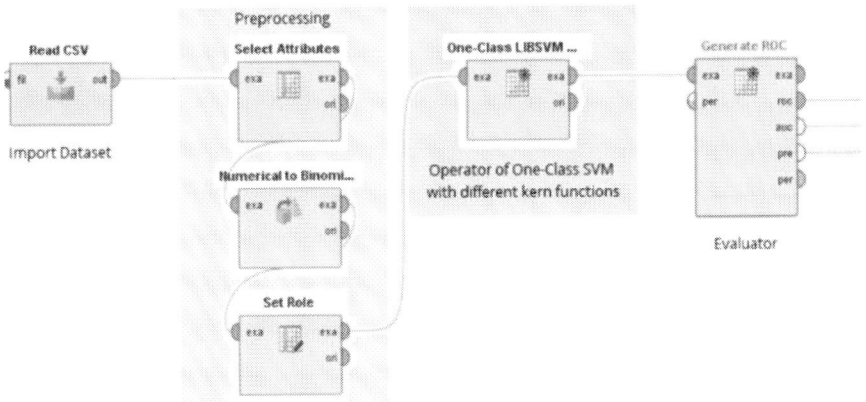

Figure 3. Application of One-Class SVM in RapidMiner.

The implemented One-Class SVM in RapidMiner includes different kernel functions like logistic, RBF and poly usw. Kern functions in this operation, while the RBF shows the best performance in this case in Table 3. In order to find the anomaly points in Japanese vowels, the result is still not efficient despite the moderate improvement of the recall ratio. The reason may lie in odd kern function or overfitting.

Table 3.

	true false	true true	class precision
pred. false	1375	31	97.80%
pred. true	31	19	38.00%
class recall	97.80%	38.00%	

The reason why the model performs poorly may be for two reasons: One possibility is the ambiguous definition of boundary. The boundary between normal and anomalous behavior is still unsearched and uncertain. For instance, through medical statistic, some rare diseases are diagnosed. However, a virus itself is not harmful to the human body, but it also has a subspecies that damages the body. Sadly, our current science does not identify a subspecies of the virus and so we mark this sample as usual, but the model detects this as outliers and vice-versa. Besides, there may be no significant difference between standard and local outliers. Worse, they are close to each other, which makes them lead to an imprecision model. Overfitting in the SVM model may mean some outliers like local outliers are also recognized as normal because some futile relationships have been found by kern function.

To conclude, One-Class SVM provides a way to solve the nonlinear issues without worry about the local minimum problem. On the contrary, due to the primary role of kern function in SVM, its explanatory is pretty vulnerable, and making them worse is the fact that they are highly sensitive to missing values.

3.8. NEURAL NETWORKS

The neural networks model is also an excellent idea for anomaly detection. Voting from hidden layers finds the best weights by forwarding feedback. Therefore, the neural networks model aims to classify with the smallest units and perceptions to approximate the original universal function. Technically, a multiplayer neural network with a certain number of units can record any assumptions about the profile of a given sufficient data.

Theoretically, the neural networks model fits any situation when there are ample perceptions. Nonetheless, the number of perceptions is uncertain. Hidden layers of neural networks should increase gradually to get the simplest perceptions. It takes time. Sometimes, just hundreds of layers are enough to reproduce the function. However, more commonly, nowadays are thousands of layers namely, deep neural network.

In conclusion, the neural networks model has high classification accuracy, and strong parallel distributed processing ability. Corresponding, the learning time is usually too long. Simultaneously, this model cannot ensure definitely ensure a good performance, even making the learning process and the results difficult to interpret.

Although in recent years, there is little room for improvement in the CPU calculating speed. Multiple computers form a cluster computing architecture, and a distributed system is also a solution to accelerate computation. Therefore, in most complicated real scenes, this type of algorithm that requires high calculation volume is also ideal.

3.9. LIMITATION OF LINEAR MODELS

Only when the Coupling of data features is high, linear models perform well. For independent scenario, although nonlinear models and kernel methods provide a valuable choice, they are computationally intensive and may often run into overfitting. Another related issue is that the fact that current correlations are incomplete. Notably, the computational complexity is an issue when the dimensionality of the data is significant, as in the neural networks model.

3.10. PROXIMITY-BASED METHODS FOR OUTLIER DETECTION

3.10.1. Introduction of Cluster-Based Model in Outlier Detection

Usually, given sufficient data contain only a part of the realistic feature. Therefore, it is tough to ensure that the dataset always has a high coupling. Sometimes, the dataset demonstrates a nonlinear shape and sometimes linear shape. As shown in Figure 1, in linear models, the locality of points is defined as a part of a curve. The definition of observations' locality from proximity-based methods is founded on proximity, as they are clustering-based, distance-based, and density-based.

In most clusters methods, they report anomalies as a side-product. A point that belongs to no one is an outlier; thus, the distance of this observation to the closest cluster centroid is the anomaly scores. Consequently, when various distance methods lead to different anomalies, i.e., Mahalanobis distance for local outliers and Euclidean Distance for global outliers.

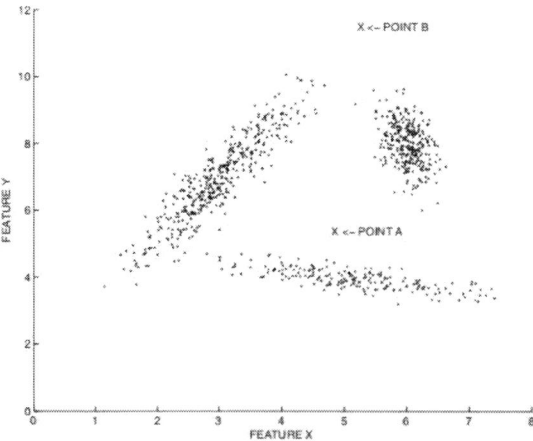

Figure 4. Location of different Anomalies.

The underlying cause of two different utility relies on how they can recognize the distance. This is evident from the example illustrated in

Figure 4. Given two outliers, A or B could be distinguished by Mahalanobis distance or Euclidean Distance. The straight line distance between A and all three clusters have no significant difference. Therefore, only global anomaly B is marked, while with Mahalanobis distance to detect, the anomaly scores of A is much more prominent than that of B. For incomplete features, the current clusters may also be not the same shape as insufficient sample size. In the extension space of other ethnic groups, not only A but also B exists (Not sure) with the same covariance matrix. Based on this reason, only local outlier A appears.

3.10.2. Clustering-based Multivariate Gaussian Outlier Score (CMGOS)

As a clustering-based method, models would arrange clusters at first. For example, in the Clustering-based Multivariate Gaussian Outlier Score (CMGOS) [4] model, the k-means method is used to identify classifications. The following covariance matrix of each cluster is also serviceable to count clusters' centric. Next, the specific method to calculate outlier's scores need different methods such as Mahalanobis Distance to recognize local outliers or with Euclidean Distance to reach global outliers. However, both distances are at the same time is available.

3.10.3. Application with Vowels Dataset

Implement this model with RapidMiner and visualize the clusters at the same time, like in Figure 5. It focuses on global outliers through exploring with Euclidean Distance. Both cluster operators in green branches are the same, the top one is to detect outliers, and the other is to visualize the clustering result.

It takes little time to detect, then gets the result in Table 4. According to the result, although the higher performance is far from accurate, recall is still a problem.

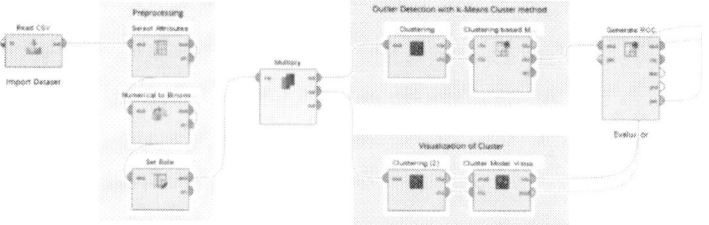

Figure 5. Application of Clustering Method in RapidMiner.

Table 4.

	true false	true true	class precision
pred. false	1378	28	98.01%
pred. true	28	22	44.00%
class recall	98.01%	44.00%	

The performance may depend on the summarization of clusters. Hence the model's accuracy and recall conjugate not only with the CMGOS model, but also with the k-means model. Hence we cannot precisely follow the source cause of fallacious. On the contrary, because of the cluster models, we can get more specific insights about the local distributions of data points.

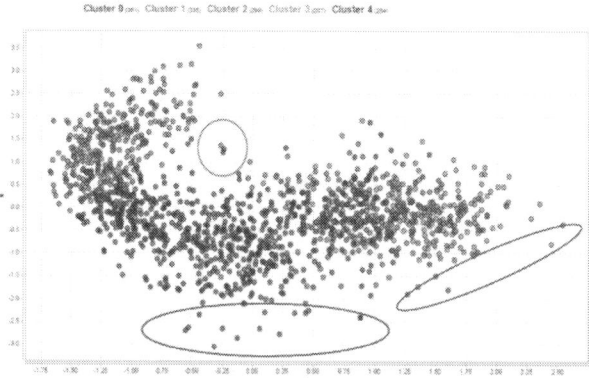

Figure 6. Distribution of Vowels Dataset.

Like in Figure 6, the visualization of data is with k-means clusters. This picture is generalized as 3-dimensional, but with cluster 2 views.

Custer 2 is obviously aroused together. As for other data points, the distribution is irregular and uneven. When this result import into CMGOS model, data points which are far away from these five clusters are outliers. In this case, only global outliers could be detected because of the used Euclidean Distance. In other words, the points like in blue circle, which are at the extreme edge may be regarded as outliers by this model so that false Negative is high in Table 4. However, the points at the center of several clusters like points in the red circle would be ignored namely, there are still many points in false positive in Table 4. In conclusion, in this case there exist not only local outliers but also global outliers. However, the created model can be found only globally so that it leads to the poor performance of recall in Table 4.

3.10.4. Distance-Based Methods for Anomaly Detection

The nearest neighbor distances are the foundation of the distance-based method; it has been generalized to all other domains such as categorical, text, time-series, and sequence data.

In comparison to the clustering-based models, the granularity of the analytical process is different, as the distance-based methods have a higher granularity to clustering-based. The samples not in clusters are potential outliers so that in Figure 7(a), only A is an outlier because this sample has no noise. However, in Figure 7 (b), every noise and observation A are all regarded as outliers clustering-based methods.

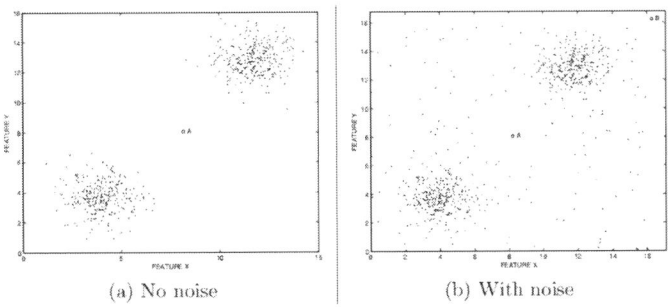

Figure 7. Effectt of noise in Anomaly Detection.

Clustering-based methods have smaller granularity than distance-based. Nevertheless, the distance-based methods know about the diversity of noise and outliers. That means, observation A and part of noises are all outliers by a distance-based method. When the parameters of the distance-based model, for instance, parameter k in the k-Nearest Neighbor algorithm, are set appropriately, all noises can be wholly eliminated.

3.10.5. Average k-Nearest Neighbor Score for Outlier Detection

The main idea of this k-Nearest Neighbor Score for outlier detection [5] is that for a given observation. Its outlier score is the average distance to its k nearest neighbors.

Similarly, we get the outlier scores of all the tested samples. Then we need to filter scores with a threshold given by an expert or given by cross-validation. There are not only mean, but also harmonic mean to normalize those k distances.

A more useful way is to give different weights to the influence of neighbors with different distances on the sample, for example to introduce the weights into the distance. In this way, the model can distinguish the difference between noise and global outliers much more preciously. At the same time, when the training sample data is in a particular space, the definition of the distance between the observation point and the neighbor is also different, for instance, *Euclidean Distance*, *Chebychev Distance, and Canberra Distance.*

3.10.6. Application with Vowels

In Table 5, the result of the average kNN methods by RapidMiner are shown. It shows a poor improvement, especially in recall ratio. As the only one parameter, k is usually fitted in an interval between 10 bis 30, decided by cross-validation. However, in this scenario, the most precious model is set by 2-nearest neighbors, like in Figure 8 by RapidMiner.

Anomaly Detection and Applications

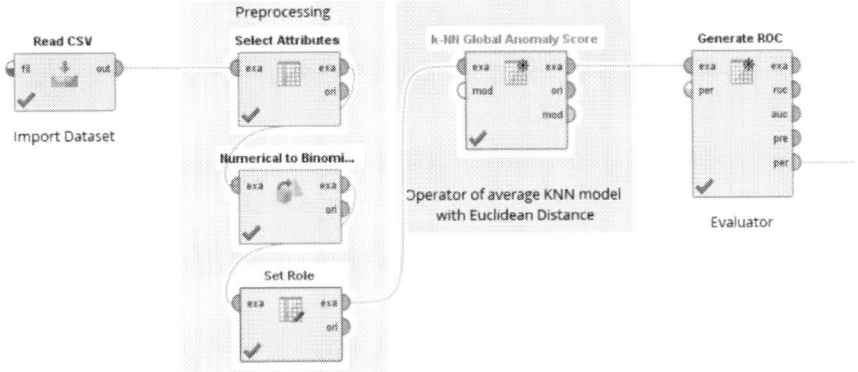

Figure 8. Application of KNN anomaly detection method in RapidMiner.

KNN model in RapidMiner is already packed with measure types and the various algorithms to get the distances. Namely, it has numerical measures with *Euclidean*, *Chebychev* and *Canberra Distance*, nominal measure with Dice, Jaccard, Kulczynski Similarity, and also Bergman divergence with Mahalanobis, Squared Euclidean, and Itakura Saito Distances. The model used in Table 5 is 2-nearest neighbors with Bregman Divergence measured by Mahalanobis Distance.

Table 5.

	true false	true true	class precision
pred. false	1390	16	98.86%
pred. true	16	34	66.00%
class recall	98.86%	68.00%	

Even though the KNN model can avoid the effect of noise theoretically, yet there are still effects from local and global outliers. Therefore, it is still possible that the model cannot fully distinguish between noise and local outliers.

3.11. DENSITY-BASED METHODS FOR ANOMALY DETECTION

Density-based outlier analysis is a further development of distance-based outlier analysis. The distance between the observation point and the kth neighbor forms a plane. The greater the density in this plane, the less likely the observation point is an outlier.

As a combination of clustering and distance-based approaches, density-based methods provide a detailed level of local insights. How data points are presented, decides varying data density rather than the varying shape and orientation.

3.11.1. Local Outlier Factor (LOF)

Local Outlier Factor (LOF) [6] shares a similar method to get the distance of points to k nearest neighbors. The given observation point o, it has a reachability distance to all other k nearest neighbors, namely the maximum distance of o to neighbors.

In order to distinguish the density of point O with its neighbors, the local reachability density of o is likewise necessary, namely, the ratio of O's average reachability, and the average of its k nearest neighbors' average reachability. However, the density of point o contains only the locality of single observation and its correlation to neighbors. Accordingly, the ratio of the density of point o to the density of its k nearest neighbors takes to be local outlier factor, specifically, the anomaly scores of LOF.

3.11.2. Application with Wisconsin-Breast Cancer Diagnostics Dataset

With this theory, more potential local anomalies would be detected successfully, like in Table 6. This model is defined by function package

CBLOF from Python without parameter judgment. This topic is about Wisconsin-Breast Cancer (Diagnostics) dataset (WBC) [7] prediction. Gratifying the recall ratios of anomalies and normal samples are all satisfactory, which are both more than 85%. Notably, the accuracy of anomalies is concurrent, not ideal, at only 32%. The reason may be too many misjudgments so that the local anomalies are detected with many other normal neighbors.

Table 6.

	true false	true true	class precision
pred. false	190	2	99.00%
pred. true	24	11	32.00%
class recall	89.00%	85.00%	

In practical applications, different demands require various levels of evaluators. For example, in Bioinformatics, to find as many potential cases as possible, relatively lower accuracy is acceptable. However, the outlier accuracy should also be higher than 33% as for the recall of potential cases is required strictly. In other words, more predicted positive observations and relatively more false predicted positives are both main optimization principles in Bioinformatics. For example, when we try to diagnose Wisconsin-Breast Cancer, after adjusting parameters, this inherited function "CBLOF" is perhaps an acceptable model.

Nevertheless, the LOF model assumes that only one point is the kth nearest neighbor. Namely, there are not many points that have the most significant distance to observation. Consequently, when the hypothesis does not hold, it may lead to errors or inaccuracies.

The high false-positive may be caused by the misunderstanding of local outliers, while the model concentrates largely on subspace to get density. It would miss many local outliers, whereas smaller subspace is associated with global outliers error.

3.11.3. Limitation of Proximity-Based Models

In Proximity-Based models, exploring global and local outliers is usually contradictory because the generation mechanisms of the two are various. It is difficult to estimate the proximity between them and the usual points by one method at the same time so that fully global analysis lose always to local outliers, whereas an entirely local analysis is prone to regard cluster in sparse spaces as global outliers. Although this issue can be solved by indexing methods to incorporate pruning into the outlier search, the effectiveness is sensitive to increasing dimensionality. Another case is for high dimensional data. The effectiveness is not eventually a fundamental limitation, but the basic definition of proximity-based model is the locality of samples. In high dimensional, some samples are equidistant, which is called the curse of dimensionality caused by data sparsity. Worse is that a small gap between samples also means a small gap between outliers and noise. Another method is necessary for high dimensional outlier detection.

3.12. HIGH-DIMENSIONAL OUTLIER DETECTION

3.12.1. Introduction of High-Dimensional Outlier detection

In association with proximity-based methods, the distances between high-dimensional points are not conspicuous in sparsity subspace, so that these models are inefficient. Concurrently, it is also challenging to extract characters because there are no full feature descriptions.

For instance, in Figure 9, the visualization from 500 samples musk dataset from the UCI machine learning repository, which has 166 features, is remarkable. The outliers concentrate on a subspace. It seems to be a simple detection. However, when we increase the sample to 2000 samples the data character in Figure 10 is unconventional. All points not only to anomalies but also to normal points range in a line. This case shows the effect of no full feature descriptions perfectly. As finite high dimensional, like in this case, the more samples we use, the more character and more noise we catch. However, we know little about how realistic are the samples we get and when and whether all characters of original data are already included.

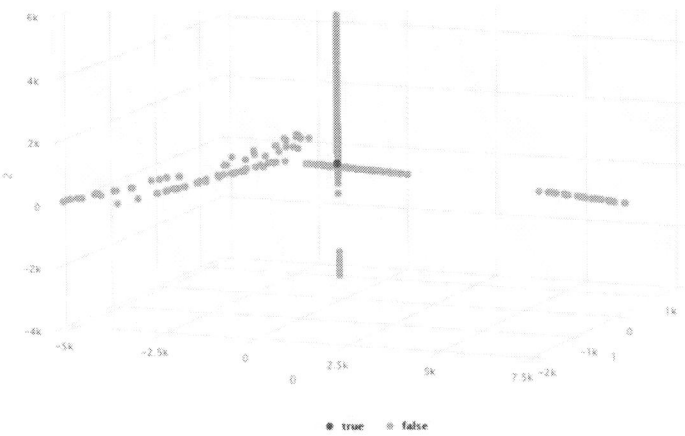

Figure 9. Anomalies in high-dimensional dataset(with 500 samples).

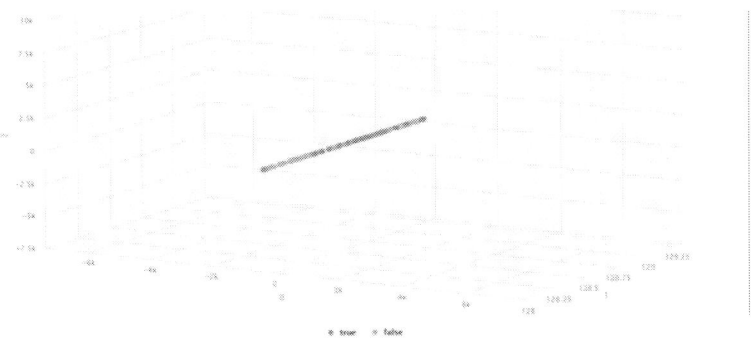

Figure 10. Details of data from high-dimensional(with 2000 samples).

Subsequently, it is executable to find proper subsets of dimensions to explore outliers. Therefore, models should be able to search the data points and dimensions in an integrated way.

Different subsets lead to distinct outliers so it is decisive to get proper subsets to improve accuracy and recall. Possible projections of high-dimensional data are rarity-based [8], unbiased [9], and aggregation-based

[10] methods. To identifying rarely populated subspaces is the main idea of rarity-based methods. As for unbiased algorithms is to sample observations in an unbiased way by feature bagging or rotated bagging. Another conception which is different from the two mentioned conceptions, but easier to understand and implement is by quantifying statistical properties.

3.12.2. Isolation Forest (IForest)

Outliers in Isolation Forest (IForest) [11] are defined in a subset of features from the original, which are much easier to isolation as regular points. Anomalies are in rarely populated subspaces by isolating.

Isolation Forest is a type of clustering ensemble model, which shares a similar idea with the Decision Tree and Random Forest. Each point in the Isolation Tree would be recursively partitioned by cutting the chosen partition with axis-parallel randomly so that each space has fewer instances until the points are isolated into singleton nodes, which contain only one instance. In the Isolation Tree model, anomalies are the points that can be isolated quickly namely, the anomaly scores are the depth of these points. When it comes to the Isolation Forest model, anomalies scores are the average depth of points.

3.12.3. Application with Wisconsin-Breast Cancer Diagnostics Dataset

Compiling with ensemble subpackage of sklearn package by IsolationForest model, relatively large anomaly scores are arranged because the model collects them with positive numbers. When we need to assemble them with many other anomaly detection methods, scores should be normalized with proper principles. Another problem in this encapsulated method is that, -1 means anomalies but not as 1 in most other models like in Table 7.

Table 7.

Using the packed IsolationForest function and predict
fromsklearn.ensemble import Isolation Forest
clf = IsolationForest()
clf.fit(x_train,y_test)
preditct = clf.predict(x_test)
Converting Result to 0-normal and 1-outliers.
convertedResult = pd.Series(preditct)
i = 0
for i in range(0,len(preditct)):
if convertedResult[i] == 1:
convertedResult[i] = 0
else:
convertedResult[i] = 1
#Evaluating
from sklearn import metrics
from sklearn.metrics import classification_report
print(metrics.classification_report(y_test, convertedResult))

After converting the scores into normalized numeric and unification label of anomalies, the performance of this model is described in Table 8. Although the eye-catching two indexes namely, recall of anomalies and precision of normal points are 100%. However, the defect of this model is also striking. The precision of the anomalies is horrible. More than ten times more points than real anomalies are marked.

The reason for inferior outliers' precision blames on the mechanism of the Isolation Forest model because the model occupies oneself with global anomalies. For the sake of high anomalies recall, all small clusters, which are far away from the main clusters, have been regarded as anomalies. The precision of normal points is also perfect consequently.

Table 8.

	true false	true true	class precision
pred. false	1067	0	100.00%
pred. true	711	60	7.78%
class recall	58.72%	100.00%	

Isolation Forest is a perfect global anomaly detection method despite ignoring all local outliers. Exploring local outliers in high-dimensional is, however, also vital.

3.12.4. Limitation of High Dimensional Method

Recently, many ensemble methods for subspace analysis were well-performed. However, the main issue remains still in changing relationships between features in subspace. To be specific, one dimension is eventually weekly relevant, and many more dimensions are locally irrelevant. This usually occurs and is still unsolved. Search further about deep generation causes of outliers may be a good way. At the same time, efficiency is another problem because exploring local subspaces to ensure robustness takes too much time. There is no other way but to increase the calculation speed of the computer. The potential solution is through deep anomaly detection techniques, according to the latest study.

3.13. INTRODUCTION OF INTEGRATED MODELS

LOF model for Wisconsin-Breast Cancer is already acceptable. However, it is not always easy to find a fitted model. Because different data-sets have distinct features and the outliers are also regularly small parts of integration. Therefore, the more frequent situation is that the accurate and recall of normal points are high, however outliers stay in low detecting accurate. Namely, the outlier recall ratio keeps at a low level and is hard to improve as in the Japanese vowels cases. Another example is recognizing the outliers of the concrete data-set from the UCI machine learning repository. A robust model should be demonstrated with complicated combinations of abundant algorithms.

3.13.1. Concrete Case

The performance of the different single model is still inefficient as in Table 9, even after cross-validation. Worse still, in this scenario, even though combing these models together is also not ideal, the anomalies recall and normal samples' accuracy stay at a low level. For improving, many other classification algorithms like Decision Tree (DR), Logistic Regression (LR), and Multi Perception Classification (MLPC) are also necessary. In this case, the author came up with a model named Stacking-based Integrated Outlier Detection (SBIOD) [12]. This model is based on five unsupervised models and cross-validation. This model is relatively complete in Figure 11. Two models would be used: The first model is comprised of five simple unsupervised base learners. The results are then input from the first integrated model into the second layer of SBIOD, the model of MLPClassifier.

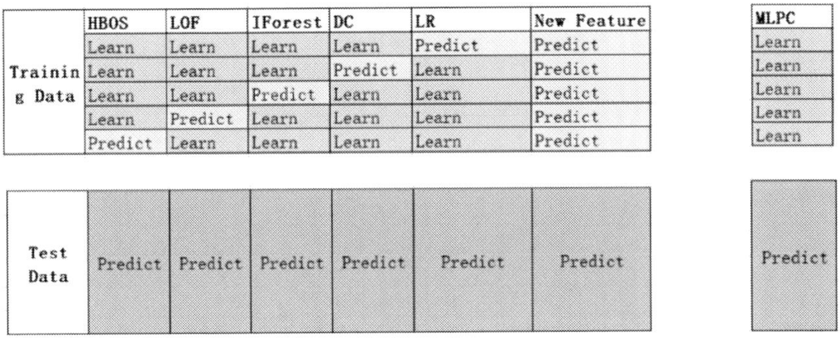

	HBOS	LOF	IForest	DC	LR	New Feature		MLPC
Training Data	Learn	Learn	Learn	Learn	Predict	Predict		Learn
	Learn	Learn	Learn	Predict	Learn	Predict		Learn
	Learn	Learn	Predict	Learn	Learn	Predict		Learn
	Learn	Predict	Learn	Learn	Learn	Predict		Learn
	Predict	Learn	Learn	Learn	Learn	Predict		Learn

Test Data	Predict	Predict	Predict	Predict	Predict	Predict		Predict

Figure 11. Structure of Stacking-based Integrated Outlier Detection.

Table 9

Collection of Concrete			
Algorithms	AUC before cross validation	AUC after cross validation	Recall
HBOS	0.286	0.302039	0.04
LOF	0.456	0.36292	0.01
IForest	0.576	0.547495	0.26
DT	0.889	0.902689	0.80

Collection of Concrete			
Algorithms	AUC before cross validation	AUC after cross validation	Recall
Logistic Regression	0.854	0.968981	0.81
MLPClassifier	0.93	-	0.87
SBIOD	-	0.976	0.98

In Table 10 AUC and anomaly are steadily increasing with the order. By using cross-validation, every single model and the integrated SBIOD model are much more reliable because of satisfied anomaly recall and AUC.

As a successful attempt, AUC, precision, and recall have significantly improved. The generality of this model is also very excellent. In Table 10, another dataset is about the details from Lublin Sugar Factory, which is used to test the integrated model. The SBIOD model is not only good at detecting anomalies in the ready-mixed concrete, but also in the manufacture of sugar.

Table 10.

Algorithms	AUC before cross validation	AUC after cross validation	Recall	Time
HBOS	0.634	0.623474	0.04	-
LOF	0.462	0.409413	0.01	-
IForest	0.534	0.479122	0.26	-
DT	0.956	0.994589	0.92	-
Logistic Regression	0.764	0.970663	0.54	-
MLPClassifier	0.5	-	0	3min
SBIOD	-	0.9828	0.94	1.3min

CONCLUSION

Analysis and applications of algorithms like Probabilistic and Statistical, linear, Proximity-Based, and High Dimensional methods are well-introduced in this chapter. The integrated example usually has higher generalization and performance. For detecting outliers, traditional anomaly detection methods are sometimes inefficient. On this account, the advanced models are critical.

Despite traditional algorithms in detecting outliers need less effort to implement, they have many scarcities. Firstly, they are sub-optimal on image and sequence datasets with complex structures. For instance, through the medical image, retinal damage is detected and discriminate musk and non-musk molecules with the complex molecular structures are exhibited. Besides, traditional methods are unavailable for data with large volumes like gigabytes. Another significant issue is the implicit boundary between normal and anomalous behavior. For example, although the white tiger is quite different from normal, as a novelty, it should be considered as normal as well. However, about species we are not familiar with, we know no boundary between normal observations and outliers. Are Samples that are significantly different from other outliers' novel?

The latest survey of outlier detection is about outliers for the video presented in Kiran et al. [2018]. Anomaly detection for Sensor Networks and the Internet of Things are also introduced by Ball et al. [2017] and by Mohammadi et al. [2017]. As for Fraud Detection, Cyber-Intrusion Detection and Medical Anomaly Detection had also progressed in 2017. Deep learning-based anomaly detection will still be there in active research, which may be extended and updated to more sophisticated techniques.

REFERENCES

[1] Billor, N., G. Kiral, A. Turkmen (2005). Outlier detection using principal components. In: *Twelfth International Conference on Statistics, Combinatorics, Mathematics and Applications,* Auburn (unpublished manuscript).

[2] Mineichi, Kudo, Jun Toyama, Masaru Shimbo. *Japanese vowels.* Information Processing Laboratory. Hokkaido University, Sapporo 060-8628, JAPAN

[3] Mennatallah, Amer, Markus Goldstein, Slim Abdennadher (2013). *Proceedings of the ACM SIGKDD Workshop on Outlier Detection and Description,* ODD '13: 8-15
[4] Zheng, Feng & Liu Quanyun. (2020). *Anomalous Telecom Customer Behavior Detection and Clustering Analysis Based on ISP's Operating Data.* IEEE Access. PP. 1-1. 10.1109/ACCESS.2020.2976898.
[5] Brito, M.R., E.L. Chávez, A.J. Quiroz, J.E. Yukich (1997). Connectivity of the mutual k-nearest-neighbor graph in clustering and outlier detection. *Statistics & Probability Letters:* Pages 33-42.
[6] Breunig, Markus M., Hans-Peter Kriegel, Raymond T. Ng, and Jörg Sander. 2000. LOF: identifying density-based local outliers. In *Proceedings of the 2000 ACM SIGMOD international conference on Management of data* (SIGMOD '00). Association for Computing Machinery, New York, NY, USA, 93–104.
[7] Dua, D., and C. Graff (2019). *UCI Machine Learning Repository* [http://archive.ics.uci.edu/ml]. Irvine, CA: University of California, School of Information and Computer Science.
[8] Dutta, J.K., B. Banerjee and C.K. Reddy, "RODS: Rarity based Outlier Detection in a Sparse Coding Framework," in *IEEE Transactions on Knowledge and Data Engineering,* vol. 28, no. 2, pp. 483-495, 1 Feb. 2016.
[9] Nguyen, H.V., V. Gopalkrishnan, I. Assent (2011). An Unbiased Distance-Based Outlier Detection Approach for High-Dimensional Data. In: Yu J.X., Kim M.H., Unland R. (eds) *Database Systems for Advanced Applications.* DASFAA 2011. Lecture Notes in Computer Science, vol 6587. Springer, Berlin, Heidelberg
[10] Sun, B., X. Shan, K. Wu and Y. Xiao, "Anomaly Detection Based Secure In-Network Aggregation for Wireless Sensor Networks," in *IEEE Systems Journal,* vol. 7, no. 1, pp. 13-25, March 2013.
[11] Liu, F. T., K. M. Ting and Z. Zhou, "Isolation Forest," *2008 Eighth IEEE International Conference on Data Mining,* Pisa, 2008, pp. 413-422.

[12] Shu, H., X. Zhao, H. Luo and C. Li, "Research on Stacking-Based Integrated Algorithm of Anomaly Detection in Production Process," *2019 International Conference on High Performance Big Data and Intelligent Systems* (HPBD&IS), Shenzhen, China, 2019, pp. 85-90.

In: Anomaly Detection
Editors: Saira Banu et al.

ISBN: 978-1-53619-264-3
© 2021 Nova Science Publishers, Inc.

Chapter 4

AN EVOLUTIONARY STUDY ON SIOT (SOCIAL INTERNET OF THINGS)

Dinesh Mavaluru[2] and Jayabrabu Ramakrishnan[2]
[1]Department of Information Technology,
College of Computing and Informatics,
Saudi Electronic University, Saudi Arabia
[2]Department of Information Technology and Security,
College of Computer Science and Information Technology,
Jazan University, Saudi Arabia

ABSTRACT

Internet of Things (IoT) is defined as the meeting point of the smart objects that communicate with another over the internet. The paradigm of the "Internet of people" is overlooked by the term "Internet of things" in recent times. There are enormous devices connected to the internet, and the figure is expected to grow exponentially with time. IoT clubs are helped by the social networking of the IoT, which in turn makes people interact with the devices over the internet. Primary importance has to be given for the challenges related to the security and privacy for a better IoT ecosystem to be built. Owing to the rising number of challenges in vulnerabilities of the devices in IoT, the minimum resources with assorted technologies along with the deficiency of proper standards naturally pave enough way for many cyber threats. This research article aims to highlight the issues that are more critical in the security of the IoT by an analysis in terms of the essential layers of the IoT and to bring out the future research of the IoT security. Last, the future challenges are discussed for the researchers to think upon and contribute towards resolving those.

Keywords: IoT, protocols, IoT security, IoT model

4.1. INTRODUCTION

In the coming fourth generations of the information age, the IoT is expected to get involved with a vast number of smart devices that are capable of computing and actuating to get connected in the World Wide Web [1][2]. The integration of the concepts in social networking in IoT paved the way for an emerging SIoT paradigm that allows the users and the smart devices to get connected and to share information [3]. Despite its plentiful technical advantages, there are enormous challenges that are faced by IoT, such as interoperability [4], factors that have an impact on security and privacy. Thanks to these challenges as they form the base for the creation of IoT –ecosystem.

Unless the challenges are solved, the SIoT will seldom reach the pinnacle of its popularity, and all it capabilities can be overshadowed. The challenges in security are mainly caused due to the unavailability of device-specific standards. Also, these smart devices have more vulnerability and give ample space for the attacks. It is known that in the fall of the year 2016, Dyn (The DNS provider that supports almost all platforms and a broad set of online services) faced a DDoS attack due to a vast number of connected IoT that were vulnerable were infected because of the Mirai (A Malware). This was reported as the largest DDoS in records [5]. Also, in the same period, the researchers identified the flaw in the well-known radio protocol ZIgbee [6]. This showed how an aerial is used for targeting a set of bulbs that are connected on the internet, helping the attackers inject a flash SOS message that gained control over the bulbs for turning it on and off. Also, these were able to replicate some malware to the nearby devices.

The final issue to be considered more seriously is the IoT privacy, specifically for the protection of sensitive data collected by IoT applications as there is a high necessity for providing complete awareness for controlling the speedy flow of the data to users automatically. Taking these challenging situations firmly, this work is done to bring out the current status and give an overview of the IoT security design. In Section II, the models that are generic for the IoT Systems with a particular interest in threats are discussed in brief. Section III discusses concepts of trust that are defined with importance for having a social relation among the entities that are not familiar. I Section IV, the definition as to how security is to be designed for supporting the paradigm of IoT by showcasing few conventional policies that need to be re-designed for addressing particular features in the world of IoT. In Section V, some of the conventional protocols that deal with security along with their solutions are analyzed in the available literature. Section VI concludes by directing the research challenges expected in the forthcoming ages and for solving the issues about IoT security.

4.2. IoT Architecture and Threats

A conventional system of IoT is usually represented in terms of five essential layers namely the a) Perception, b) Network c) Middleware d) Application, and the Business layers. Figure 1 elucidates the levels of these systems along with the technologies adopted which bring out the anticipated weaknesses in the security of IoT. The different issues in security are analyzed in [7] with the hope of finding new solutions that are more feasible and robust.

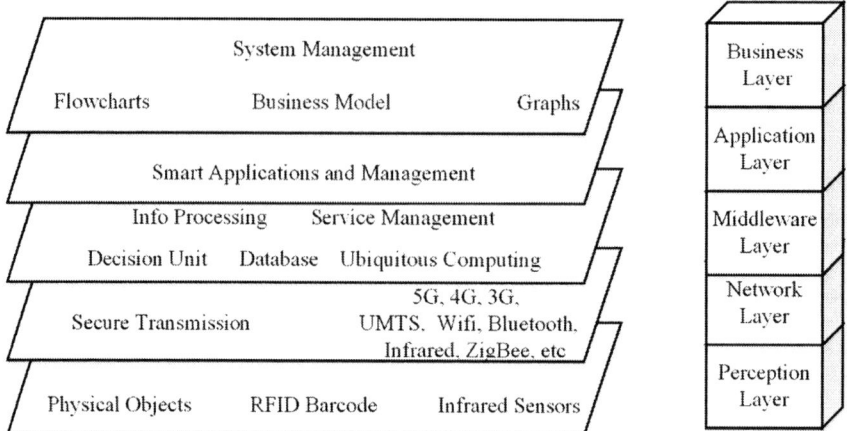

Figure 1. Layers of IoT Architecture.

A. Perception Layer

The first layer of an IoT model is the perception layer, which is involved with the sensors in IoT devices for collecting data, and processing is carried out through technologies such as RFID, WSN, RSN, and GPS. The physical layer includes different sensors for observing measurements such as temperature, humidity, pressure, altitude, etc. and also performs location identification functionalities. [8]. the nodes here are limited to the resources and are also subjected to have a distributive structure in their organization and have the following security menaces in the physical layer.

B. Transport Layer

This layer ensures omnipresence access to the initial level. The main task of this level is for the transmission of the collected data that are obtained from the first level to another system for processing. The collected information is distributed over any network that is used by all access networks (4G, WiFi, MANET) or through the internet. A general view of the security challenges in a cellular wireless network is given in

[9]. Based on this paper, the different and open frameworks in an LTE network, which are customarily IP-based, are subjected to more vulnerability in security when compared with the other generations of cellular networks.

C. Application Layer

These are layers that meet the customers' requirements. For example, this layer can provide the customers with data such as atmospheric pressure, temperature and other such measures when queried. This layer is essential for IoT development as it serves as a ground that portrays the different needs of the customers to develop multi-purpose IoT devices. Various environments of IoT can be developed and implemented in this layer. The applications are subjected to support its sub-level in all the services provided for a better realization of intelligent computing.

4.3. TAXONOMY OF TRUST IN IoT

Trust does not get confined with a single perception or domain. It is prevalent across different domains, and it is multidimensional and has great importance in almost all aspects of living [10]. According to the cook [11], it is defined as the degree to which the behavior of a specific entity is believed. In the IoT perspective, there is a high essentiality for the development of trust paradigms to help the consumers to prevail over the uncertainty and the potential risks involved in using the services provided over IoT [12] [13] [14]. In a specific context of SIoT, trust has a significant role in developing a social relation among the entities that are not known to each other. In reality, in this situation, the IoT devices imitate the social mannerism of their human counters independently as per the owners' networking, and they build a relation among other devices that are trusted to give services for the users.

A. Properties of Trust

In a generic context, the Trust must attain specific essential properties for the deployment in an IoT atmosphere.

B. Trust Management

Trust building and managing techniques are primarily useful for ensuring security and as a result, have been made use of in numerous applications that include joint Web-based forums [15], social networking [16], Web semantics [17], or e-Commerce [18]. For the trust to be managed across the various IoT devices, a handful of security issues have to be resolved [19] [20].

4.4. IoT Security

The aspect of security is either neglected or treated as a post-production process by the manufacturers of IoT devices. This mindset is mainly due to the driving force to achieve a shorter time to reach the market and low-cost manufacturing, leaving behind the security issues not addressed. There are only a handful of devices that give a meagre protection by implementing software-based solutions. As a result of having a high focus on software-based protection, the hardware is left vulnerable intended to allow attacks. It is evident from work [21] that if the hardware is not secured, it will invariably pave the way for the software stake also to become insecure. This section discusses the design of techniques for ensuring security in IoT devices and also gives clarity on the difference it has with conventional information technology security.

A. Fundamental Aims of Security: CIA Model

The CIA model uses the triad as put in Figure 2 for developing any methods for ensuring security by giving primary attention to three critical areas that are as follows:

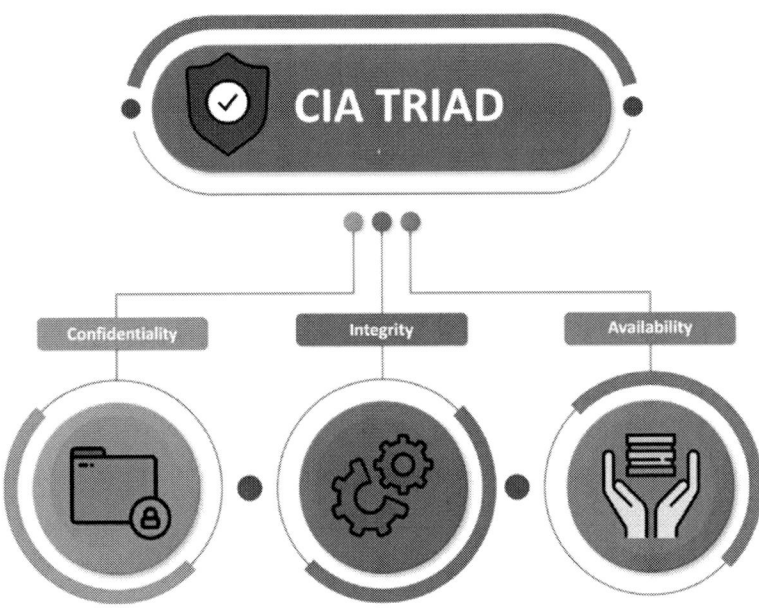

Figure 2. CIA model for data security.

Data confidentiality is the way of providing confidence to the end-user on the privacy he/she owns of the compassionate and essential information by various methods so that the disclosure to the third party, who does not have the required authorization is stopped. These are achieved through techniques such as encryption and by having suitable control over the accesses.

Data integrity is the protection of sensitive information from the attackers or hackers when the data are transmitted or when at rest by implementing suitable integrity mechanisms that stops alterations in the data.

Data availability assures that the data are available at any point of time to the authorized user, both in at reasonable time and at the time of disaster. DoS type of attacks can reduce the availability of data to authorized users. These are safeguarded using mechanisms such as the implementation of firewall policies and suitable Intrusion Detection Systems (IDS)

B. Conventional Security in IT versus IoT Security

The fundamental problem with most of the IoT devices is that they have a closed feature. There are only limited chances for the consumers to make any add-ons in the security post-production. Owing to this reason, the IoT devices are supposed to have enough in-built security (In the manufacturing process itself). In other means, the security shall be provided as an add-on, where the customers add security in the prevailing system like PCs or Mobiles (Conventional IT).

One of the significant other issues in the IoT is that the devices are manufactured with limited resources and limited hardware specifications (Ex: RFID nodes), whereas the conventional IT devices are built with more resources. Hence, only light algorithms can be deployed and also it is essential to bridge the gap between high security and lower performance.

Also, the heterogeneous nature of the IoT devices can be sensed very quickly in almost all the elements such as identifying, communicating, computing and servicing [22]. In the future, there are high chances for a range of things that can be connected to the Internet, starting from TV remote till airships. All these devices generate a massive amount of data that cannot be managed [23]–[25]. The cons of the security are kept in line with the increase in the attack. Various techniques that are heterogeneous are coupled along with their respective issues also bring in potential weakness in the security.

On the other hand, as far as the application layer is concerned, the issues about privacy have more challenges as the IoT-based applications are used for almost all the day-to-day works, and our private information is being captured continuously for making our life simple and hassle-free. These IoT devices have data that can even control our living environment and can bring out a significant security issue once we happen to lose control over them. The lack of additional security makes conventional IT security less vulnerable than IoT security.

To summarize, the IoT applications and systems are deployed in an environment that is more dangerous and vulnerable, with their resources limited along with a constraint of minimum security. Hence, fewer weight solutions have to be developed for handling these vulnerable atmospheres, where the possibilities for attacks are more. Table 3 gives the fundamental differences in conventional IT security with that of the IoT security in the context of requirements and application.

Table 1. Conventional security versus IoT security

Conventional IT security	IoT Security
Security as an Add-on	Security is in-built
Complex and heavy methodology	Simple and lightweight methods
User has a controlling hand on the security level	No privacy preserved – Most data are collected automatically
Minimum heterogeneity	Large heterogeneity
Devices will be in "Closed" atmosphere	Devices will be in an "open" atmosphere.

4.5. ISSUES AND SECURITY SOLUTIONS FOR IoT COMMUNICATION PROTOCOLS

The primary step for the inclusion of security in the IoT devices is also related to the secured communication that uses a variety of protocols that are used in such a way that the transmission of data always follows the CIA mandate. In the point protocols, the IoT-based protocols can be segmented into three categories. [26]. *Physical accessibility, Networking, and the Application.* This section tabulates the commonly used protocols in an IoT system, along with the expected threat and the solutions provided so far. It is given as an open forum for the researchers to contribute to enhancing these protocols or to introduce new protocols that are needed for the hour.

A. Protocols, Issues, and Solution in SIoT

Some of the protocols that are available for the security of IoT and the current issues along with the solutions are tabulated below and are self-explanatory.

Table 2. Protocols, issues and solutions

	Protocol Used	Issues	Solution
Physical Layer	IEEE 802.15.4	Potential attacks while data transmission	CEM algorithm [28]
	BLE	Non-encrypted headers	Black Network [28]
	WI-Fi	Transmission Attacks	WEP, WPA protocols [29]
Network Layer	IPv4/v6	NDP Protocol Threats	SEND [30]
	LoWPAN	Transition Attacks	DTLS [31]
	RPL	Routing Attacks	SVELTE [32]
Transport layer	MQTT	Nonscalable management of keys	Secure MQTT with ABE [33]
	COAP	High Computation Cost	Lithe [34]

B. General Security Issues in IoT

The concept of security has always been a primary challenge in the widely spread adaptation in any technology. As far as SIoT is considered, the concept of trust and security is more essential for the development of any relationship that is mutual between the devices. The protocols in the OS must be robust and flexible in order to ensure the authentication and trust-building tasks that can carry out efficient applications. Various types of protocols are used to ensure that the IoT is secured. Also, many algorithms are developed for the sake of enhancing security, but these algorithms must be less expensive and also take the time consumption should be considered as an essential factor. These are also expected to work in a real-time scenario for avoiding any degradation in performance. The improper handling of memory also increases security vulnerability [27].

4.6. CHALLENGES IN IOT - NEAR FUTURE

There is no doubt that the digital worlds are ruled by the smart devices. The forth coming generations will have numerous connected devices. In spite of having many advantages of having devices connected over the internet, these are also subjected to have some drawbacks or challenges in the future and the same are discussed below.

4.6.1. Lack of Security

Given the choice to the manufacturers for selecting either convenience or security, the choice will be to convenience as per the producers. The IoT devices are no exceptions in this case. Most of the devices that are connected in a network are vulnerable to hacking and this was reported in 2015 when a car that was self-driving was hacked. Crucial information is overloaded every second in the internet and hence security plays a vital role. The concept of human-driven IoT devices should be maintained at all times of bringing out IoT-driven humans [28].

4.6.2. Lack of Privacy

Invariably, the devices in the IoT are subjected to have a lot of sensitive data of a user and can be subjected to vulnerability and can be attacked in any form over the network. The smart devices can be kept on tracking and the information is liable to get leaked. Hacking one's smart phone may result in access to his vital and confidential information. Hence, even there is large number of privacy preserving techniques for the SIoT [29], there still seems to be more vulnerability and open research challenges still persist in these areas.

4.6.3. Storage Issues

The amount of data that are available in the internet if printed will take the distance from planet earth to mars. There data that gets generated by the smart devices will surely increase in the near future. The data available in the [30] internet are expected to get doubled every 5 years. Hence, suitable mechanisms are needed to handle the growing data generated by IoT devices.

4.6.4. E-Waste

Survey has revealed that the e-waste by the scrap of electronic devices in USA alone is fifty million tons a year. Also, many developing countries are advancing economically and hence the production of e-waste [31] likely to exponentially. Only a meager portion of the wastes are re-cycled and the others are left unattended.

4.6.5. Energy Demands

As per the prediction by Gatner, around 6.4 billion devices would be used and it will obviously be more with the time. Hence, there would be a great demand for the energy. In the year 2018, it was estimated that the internet working alone needs 90 billion watts of power for a year [32].

Thus new technologies have to be designed for the efficient energy usage or less-powered devices.

4.7. CONCLUSION

To get along with the fast development in the IoT sector, the security importance is also growing exponentially. This chapter has shown that the models used so far have some security issues to be addressed and to exploit some of the potential weaknesses. It is for a similar reason that there is an urgent need for the enforcement of trust and security in the world of IoT, ranging from the basic models to those of higher level as various threats are related to different IoT models. In this chapter, the perception layer is identified as most vulnerable owing to its heterogeneous nature and also the fact that these are subjected to get exposed physically. Thus, it is crucial, in the next future, to start working on the critical issues of this level implementing lightweight security solutions that can adapt to the heterogeneous environments with resource-constrained devices. The expected challenges shortly are also discussed that throws some light on the researchers to make a decent contribution in resolving those.

REFERENCES

[1] Karnouskos, S; Marrón, PJ; Fortino, G; Mottola, L; Martínez-de Dios, JR. *Applications and Markets for Cooperating Objects* (Springer Briefs in Electrical and Computer Engineering). Heidelberg, Germany: Springer, 2014, pp. 1–120.

[2] Fortino, G; Trunfio, P. *Internet of Things Based on Smart Objects, Technology, Middleware and Applications*. Cham, Switzerland: Springer, 2014.

[3] Xiao, H; Sidhu, N; Christianson, B. "Guarantor and reputation based trust model for social Internet of Things," in *Proc. Int. Wireless*

Commun. Mobile Comput. Conf. (IWCMC), Dubrovnik, Croatia, 2015, pp. 600–605.

[4] *Inter-IoT Project*. Accessed: Oct. 2017. [Online]. Available: http://www.inter-iot-project.eu/.

[5] Wikipedia Contributors. (2018). *Dyn Cyberattack*. [Online]. Available: https://en.wikipedia.org/w/index.php?title=2016_Dyn_cyberattack&oldid=763071700.

[6] Ronen, E; Shamir, A; Weingarten, AO; O'Flynn, C. "IoT goes nuclear: Creating a ZigBee chain reaction," in *Proc. IEEE Symp. Security Privacy (SP)*, San Jose, CA, USA, 2017, pp. 195–212.

[7] Jing, Q; Vasilakos, AV; Wan, J; Lu, J; Qiu, D. "Security of the Internet of Things: Perspectives and challenges," *Wireless Netw.*, vol. 20, no. 8, pp. 2481–2501, Nov. 2014.

[8] Lin, K; Chen, M; Deng, J; Hassan, MM; Fortino, G. "Enhanced fingerprinting and trajectory prediction for IoT localization in smart buildings," *IEEE Trans. Autom. Sci. Eng.*, vol. 13, no. 3, pp. 1294–1307, Jul. 2016.

[9] Barakoviˊc, S; et al., "Security issues in wireless networks: An overview," in *Proc. XI Int. Symp. Telecommun. (BIHTEL)*, Sarajevo, Bosnia and Herzegovina, 2016, pp. 1–6.

[10] Wanita, S; Surya, N; Cecile, P. "A survey of trust in social networks," *ACM Computing Survey*, vol. 45, no. 4, pp. 1-33, Aug. 2013.

[11] Cook, S. "Trust in Society," *Russell Sage Foundation Series on Trust*, vol. 2, Feb. 2003.

[12] Pinto, S; Gomes, T; Pereira, J; Cabral, J; Tavares, A. "IIoTEED: An enhanced, trusted execution environment for industrial IoT edge devices," *IEEE Internet Comput.*, vol. 21, no. 1, pp. 40–47, Jan./Feb. 2017.

[13] He, D; Chen, C; Chan, S; Bu, J; Vasilakos, AV. "ReTrust: Attack-resistant and lightweight trust management for medical sensor networks," *IEEE Trans. Inf. Technol. Biomed.*, vol. 16, no. 4, pp. 623–632, Jul. 2012.

[14] Kounelis, I; et al., "Building trust in the human–Internet of Things relationship," *IEEE Technol. Soc. Mag.*, vol. 33, no. 4, pp. 73–80, Nov. 2014.

[15] Cho, JH; Swami, A; Chen, R. "A survey on trust management for mobile ad hoc networks," *IEEE Communications Surveys and Tutorials*, 13(4), pp. 562-583, 2011.

[16] De Meo, P; Musial-Gabrys, K; Rosaci, D; Sarnè, GML; Aroyo, L. "Using centrality measures to predict helpfulness-based reputation in trust networks," *ACM Trans. Internet Technol.*, vol. 17, no. 1, pp. 1–20, 2017.

[17] Lin, WY; Zhang, X; Song, H; Omori, K. "Health information seeking in the Web 2.0 age: Trust in social media, uncertainty reduction, and self-disclosure," *Comput. Human Behav.*, vol. 56, pp. 289–294, Mar. 2016.

[18] Shirgahi, H; Mohsenzadeh, M; Javadi, HHS. "Trust estimation of the semantic Web using semantic Web clustering," *J. Exp. Theor. Artif. Intell.*, vol. 29, no. 3, pp. 537–556, 2017.

[19] Vijay, TS; Prashar, S; Parsad, C. "Online shoppers' satisfaction: The impact of shopping values, website factors and trust," *Int. J. Strategic Decis. Sci.*, vol. 8, no. 2, pp. 52–69, 2017.

[20] Lacuesta, R; Navarro, G; Cetina, C; Penalver, L; Lloret, J. "Internet of things: where to be is to trust," *EURASIP Journal on Wireless Communications and Networking*, 2012(1), pp. 1-16, 2012.

[21] Yan, Z; Zhang, P; Vasilakos, AV. "A survey on trust management for Internet of Things," *Journal of Network and Computer Applications*, vol. 42, pp. 120-134, Jan. 2014.

[22] Arias, O; Wurm, J; Hoang, K; Jin, Y. "Privacy and security in Internet of Things and wearable devices," *IEEE Trans. Multi-Scale Comput. Syst.*, vol. 1, no. 2, pp. 99–109, Apr./Jun. 2015.

[23] Al-Fuqaha, A; Guizani, M; Mohammadi, M; Aledhari, M; Ayyash, M. "Internet of Things: A survey on enabling technologies, protocols, and applications," *IEEE Commun. Surveys Tuts.*, vol. 17, no. 4, pp. 2347–2376, 4th Quart., 2015.

[24] Xu, X; Ansari, R; Khokhar, A; Vasilakos, AV. "Hierarchical data aggregation using compressive sensing (HDACS) in WSNs," *ACM Trans. Sensor Netw.*, vol. 11, no. 3, 2015, Art. no. 45.

[25] Qin, Y; et al., "When things matter: A survey on data-centric Internet of Things," *J. Netw. Comput. Appl.*, vol. 64, pp. 137–153, Apr. 2016.

[26] Wan, J; et al., "Software-defined industrial Internet of Things in the context of industry 4.0," *IEEE Sensors J.*, vol. 16, no. 20, pp. 7373–7380, Oct. 2016.

[27] Aloi, G; et al., "A mobile multi-technology gateway to enable IoT interoperability," in *Proc. IEEE 1st Int. Conf. Internet Things Design Implement. (IoTDI)*, Berlin, Germany, 2016, pp. 259–264.

[28] Sheng, Z; et al., "A survey on the IETF protocol suite for the Internet of Things: Standards, challenges, and opportunities," *IEEE Wireless Commun.*, vol. 20, no. 6, pp. 91–98, Dec. 2013.

[29] Chakrabarty, S; Engels, DW. "Black networks for Bluetooth low energy," in *Proc. IEEE Int. Conf. Consum. Electron. (ICCE)*, Las Vegas, NV, USA, 2016, pp. 11–14.

[30] Adnan, AH; et al., "A comparative study of WLAN security protocols: WPA, WPA2," in *Proc. Int. Conf. Adv. Elect. Eng. (ICAEE)*, Dhaka, Bangladesh, 2015, pp. 165–169.

[31] Hennebert, C; Santos, JD. "Security protocols and privacy issues into 6LoWPAN stack: A synthesis," *IEEE Internet Things J.*, vol. 1, no. 5, pp. 384–398, Oct. 2014.

[32] (Dec. 2014). *Bluetooth Core Version 4.2*. [Online]. Available: https://www.bluetooth.com/specifications/adopted-specifications.

In: Anomaly Detection
Editors: Saira Banu et al.
ISBN: 978-1-53619-264-3
© 2021 Nova Science Publishers, Inc.

Chapter 5

A CRITICAL STUDY ON ADVANCED MACHINE LEARNING CLASSIFICATION OF HUMAN EMOTIONAL STATE RECOGNITION USING FACIAL EXPRESSIONS

Jayabrabu Ramakrishnan[1,], PhD*
and Dinesh Mavaluru[2], PhD

[1]Department of Information Technology and Security, College of Computer Science and Information Technology, Jazan University, Saudi Arabia
[2]Department of Information Technology, College of Computing and Informatics, Saudi Electronic University, Saudi Arabia

---

[*] Corresponding Author's E-mail: jayabrabu@gmail.com.

ABSTRACT

Facial expressions are the main features of non-verbal communication. Facial Expression Recognition [FER] is the most required and one of the most challenging techniques in social communication. This chapter proposed to review the related works done on facial expression recognition. This chapter examines the basics of FER techniques, which include the three major stages such as pre-processing, feature extraction, and classification. In this survey, a comparison of various classifiers, for example, KNN, LDA, ANN, HMM, SVM, and CNN, were implemented to the FER system is also discussed. The accuracy of different classifiers techniques concerning different datasets is discussed in this survey. This study concludes on the recognition accuracy as the output parameter, and that the deep learning-based classifiers performed the best recognition result as compared to the other types of classifiers. The fundamental goal of this chapter is to provide the summarized result about profound learning-based classifiers benefits, advantages, and overall accuracy rate among the other techniques as well as providing future research directions for other potential reviewers.

Keywords: facial expressions, machine learning, human emotional state recognition, deep learning classifier

5.1. INTRODUCTION

Facial expression is a standout amongst the most dominant, global, and natural signals for humans to communicate their emotions and expectations. Many researchers have been directed on automated facial expression detection due to its practical significance in medical treatment, sociable robotics, driver weariness reconnaissance, and numerous other human-computer association frameworks. In the space of computer vision and ML, different FER frameworks have been developed to encode expression data from the representation of facial [7]. FER is a dynamic area of research. It is an intriguing segment of image processing [5].

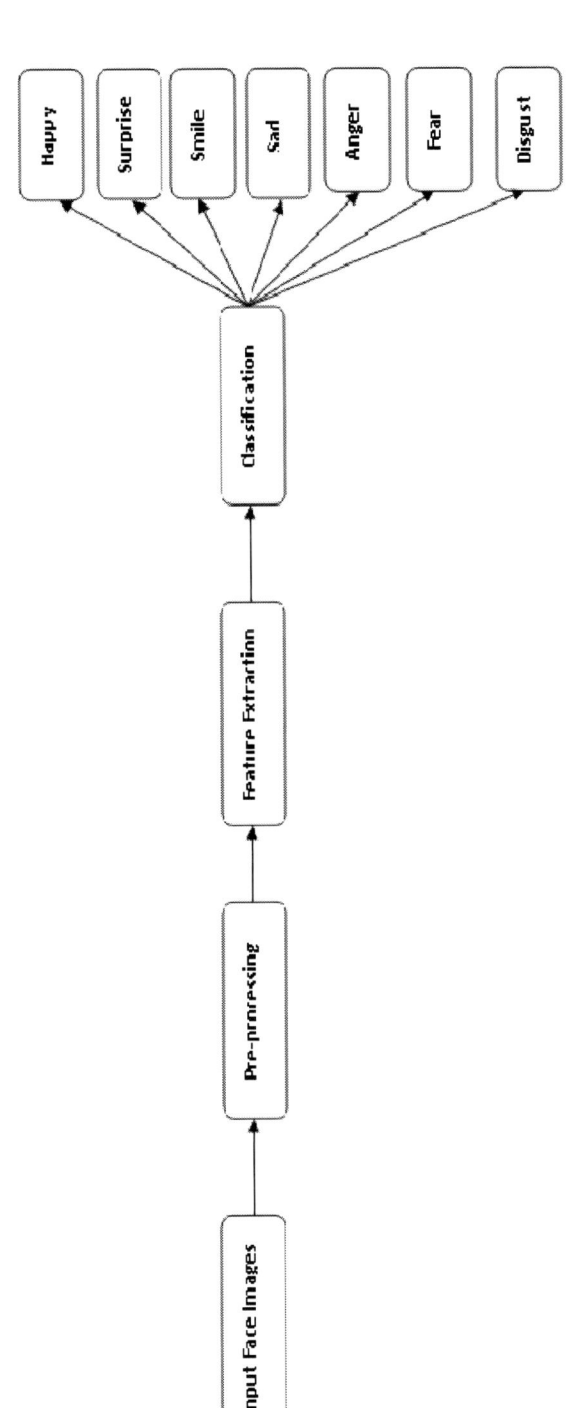

Figure 1. Representation of the FER system.

The method of identification and recognition is significant because of the comparativeness of facial expressions. The distortion happens because of emotions expressed on the faces. Despite the difference in age, gender, ethnicity, there is a resemblance in facial expressions. Darwin first reported it in 1872 and 100 years later, Ekman and Friesen proposed six expressions known as happiness, sadness, surprise, anger, disgust, and fear. These six expressions are known as fundamental emotions with their own and unique nature. This common likeness in facial expressions of humans is used by each FER framework [12]. An extensive range of processes has proposed distinguishing expressions such as happiness, sadness, surprise, anger, disgust, and fear, yet others are hard to be carried out [10, 92].

The FER framework comprises the main stages like face image pre-processing, feature extraction, and classification. In this chapter, the first section discusses the basic techniques and requirements of FER and deep learning methods. The second section discusses major reviews on Facial Expression Recognition from 2003 to 2019 years articles from standard publishers like IEEE, Springer, and Elsevier. The third section discusses the classifiers based on FER techniques in which more than 22 research articles IEEE, Springer, and Elsevier for the period of 2006 to 2019 are to compare the classifier techniques for FER. In the fourth and fifth sections, the discussion is about feature-based FER and deep learning-based FER methods and its techniques. The final section discusses the performance of different deep learning classifier techniques of FER concerning datasets and classifier algorithms, its accuracy rate. Table 4 of this chapter deals with the advantages of an in-depth comparison of learning classifiers. It clearly shows why the author considered and concentrated his research on deep learning classifier techniques. Finally, the conclusion and suggestions are discussed towards the end of the chapter.

5.1.1. Pre-Processing

Pre-processing is a method that can be utilized to enhance the exhibition of the FER framework, and it very well may be done before the feature extraction method. Image pre-processing comprises of various procedures like the clarity of image and scaling, contrast alterations, and more improvement procedures to enhance the expression outlines. The scaling and cropping methods were made on the face image in which the nose of the face is considered the midpoint, and the various significant facial segments are comprised physically.

Bessel down-sampling is utilized for face image reduction of size. However, it ensures the viewpoints and the perceptual value of the actual image. The Gaussian filter is utilized for resizing the input images that give the smoothness to the images. Normalization is the pre-processing technique that can be intended for the decrease of brightness and modifications of the face images with the median filter and to accomplish an enhanced image. The normalization technique additionally utilized for the extrication of eye positions that make it increasingly hard to identity contrasts for the FER framework, and it gives greater clarity to the input images. The localization is a pre-processing technique, and it utilizes the algorithm called Viola-Jones to distinguish the facial images from the input. Recognition of size and position of the face images are done utilizing the algorithm AdaBoost learning and Hear-like features.

The localization is mainly utilized for recognizing the size and positions of the face from the image. Face arrangement is additionally the pre-processing technique that can be executed by utilizing the SIFT flow algorithm. For this, initially the reference image for every expression of face is computed. From that point onward, every image is ordered by related reference images. ROI segmentation is one of the significant sorts of pre-processing technique that comprises of three significant functions like directing the face measurements by partitioning the color segments and of the face image, mouth region, and eye or forehead segmentation. In FER, segmentation of ROI is most familiar due to the convenient segmentation of face organs from the face images. The histogram equalization technique is utilized to vanquish the variations in brightness. This strategy is majorly utilized for improving the contrast of the images and for actual lighting additionally utilized to enhance the difference among the intensity.

In the FER, many pre-processing techniques are utilized. However, the segmentation of the ROI process is progressively reasonable because it recognizes the facial parts precisely, which parts are mostly utilized for expression detection. Next, the histogram equalization is additionally another significant pre-processing procedure for FER because it enhances the image quality [4].

5.1.2. Feature Extraction

Feature extraction changes over pixel information into a more elevated representation of color, motion, shape, spatial, and texture design of the face or its parts. Feature extraction, for the most part, decreases the dimensionality of the space of input. The reduction technique must hold fundamental data as it is a significant undertaking in the pattern detection framework. Feature extraction should be possible, utilizing different methods [10]. The extraction of the feature process is the following phase of the FER framework. Feature extraction is detecting and representation of positive features of consideration inside an image for additional processing. In computer vision image processing of feature, extraction is a critical stage, though it detects the move from graphic to specific information representation. At that point, this information delineation can be utilized as a contribution to the classification. The extraction of feature techniques is classified into five sorts: edge-based method, texture feature-based method, geometric feature-based method, patch-based method, and global and local feature-based method. Extricating the best features is a standout amongst the most significant strides of any promising FER framework.

The efficiency and viability of the facial image portrayal could impact the solidness during the process of recognition. In building up a precise FER framework, feature extraction is the essential stage. Unprocessed facial images hold large measures of information, and feature extrication is needed to reduce it to small sets of information called features. Feature extrication changes pixel data into an increasingly real portrayal of motion, color, shape, and spatial setup of the face or its features. The isolated representation is used for additional expression classification. Feature extraction conventionally diminishes the data's dimensionality space [15].

5.1.3. Classification

Expression classification is performed by a classifier, which frequently comprises of models of pattern dissemination, coupled to a decision system. Ekman characterized two primary sorts of classes utilized in FER those are AUs and prototypic facial expressions. Different classification techniques are utilized to extricate expressions [10]. Classification is the last phase of the FER framework, where the classifier classifies the expression [4]. The classification relates to an algorithmic methodology for perceiving a given expression as a given number of expressions [15]. Various classifiers, for example, KNN, LDA, ANN, HMM, SVM, and CNN, can be implemented to the FER framework. A broad range of classifiers, including parametric, just as nonparametric methods, has been utilized to the FER issue [14].

The two fundamental kinds of classes utilized in FER are activity units (AUs) and the prototypic facial expressions characterized by Ekman. In any case, it has been noticed that the changes in complexity and significance of expressions include more than these six expression classes. Additionally, numerous exploratory expression recognition frameworks utilize prototypic expressions as output classes. Such expressions happen inconsistently and subtle modifications in one or a couple of discrete facial parts impart expressions and intentions. An AU is one of 46 atomic components of prominent facial movement or its related deformation: an expression results typically from the agglomeration of many AU. AUs are portrayed in the Facial Action Coding System (FACS). Occasionally, AU and prototypic expression classes are both utilized in a hierarchical recognition framework. For example, classification into AUs can be utilized as a low dimension of expression classification, trailed by high-level AU classification integration into fundamental expression models.

5.1.4. Facial Expression Databases

Facial expression existing databases are ordered into two classes: Lab-based database, where the emotions are deliberately expression under a controlled condition, and sensible database, where the emotions frequently happen in an uncontrolled condition (for example, functional conditions). Most of the current facial expression database has a place with the top of the line, which comprises JAFFE, BU-3DFE, CK+, Simazine, SAL, MMI, AAI, and NVE. As opposed to lab-based emotions, the sensible facial expression contains massive changes in brightness, face pose, size, and in this way, they are more testing to classes and have more significance in practical applications [3]. Some of the existing databases are described in the following table. In the table.1 *P is indicating Posed, and S is indicating Spontaneous.*

Table 1. Comparative study of databases used for FER.

Database	Samples	Subject	Collection Condition	Elicitation Method	Expression distribution
CK+	593 images Sequences	123	Lab	P & S	6 basic expressions plus contempt and neutral
JAFFE	213 images	10	Lab	P	6 basic expressions plus neutral
MMI	740 images and 2,900 videos	25	Lab	P	6 basic expressions plus neutral
TFD	112,234 Images	N/A	Lab	P	6 basic expressions plus neutral
FER-2013	35,887 images	N/A	Web	P & S	6 basic expressions plus neutral
SFEW 7.0	1,766 images	N/A	Movie	P & S	6 basic expressions plus neutral
Motioned	1,000,000 Images	N/A	Web	P & S	23 basic expressions or compound expressions
RAF-DB	29,672 images	N/A	Web	P & S	6 basic expressions plus neutral and 12 compound expressions

There are a few variations among FER and face detection. Face detection is to verify which a specific image has a face [10]; we should be most likely to characterize the standard structure of the face.

Table 2. Differences between FER and Face Recognition

Facial Expression Recognition	Face Recognition
It is computer software for identifying the facial expressions of any person, one or the other using an image or a video clip or the person himself/herself.	It is a computer application for automatically identifying or verifying a person from a digital image or a video frame.
Procedurals steps: Pre-processing, feature extraction, and expression classification.	Procedurals steps: Data acquisition, input processing, faces image classification, and decision making.
Applications: Health care, games, E-learning.	Applications: Voter verification, banking using ATM, mobile password.

5.1.5. Deep Learning

Deep learning is many learning strategies endeavoring to demonstrate information with complex structures consolidating various nonlinear changes. The basic blocks of deep learning are the neural networks that are joined to form deep neural networks [11, 28, 33, 62]. These systems have empowered massive advancement in sound and image processing, including facial detection, speech recognition, PC vision, automatic language processing, classification of text (for instance, spam detection), and numerous different domains, for example, drug detection and genomics. Potential applications are varied.

Deep learning permits computational models that are made from various processing layers to learn a description of information with many dimensions of abstraction [44, 95]. Deep learning finds unpredictable structures in extensive datasets by utilizing the backpropagation algorithm to show how a machine should change its interior parameters that are utilized to process the demonstration in each layer from the representation in the past layer. The deep convolutional network has realized leaps forward in processing video, image, audio, and speech, though recurrent networks have shone a light on the following information, for example, speech and text [17]. Deep learning is generally executed utilizing the architecture of the neural network. The expression "deep" alludes to the total layers in the network - the more layers, the more profound the network. Conventional neural networks contain just 2 or 3 layers, while deep networks can have hundreds. Here are some of the profound learning examples at work:

- A self-driving vehicle gets slow as it approaches a pedestrian crosswalk.
- An ATM rejects a fake banknote.
- A cell phone application presents a quick interpretation of a foreign sign or language.

Deep learning is exceptional because of its accuracy. Advanced tools and strategies have significantly enhanced deep learning algorithms to the point where they can overcome the performance of a human [17, 95, 11, 28, 44, 33, 62].

Three innovation empowering influences make this level of accuracy conceivable:

- Easy access to large sets of labeled information
- Expanded computing power.
- Pre-trained models developed by specialists.

A deep neural network consolidates various nonlinear processing layers, utilizing essential components performing simultaneously and stimulated by biological nervous systems. It comprises an input layer, many hidden layers, and the output layer. The layers are interlinked utilizing nodes, or neurons, with each hidden layer utilizing the output of the last layer as its input. There exist many sorts of neural network architectures:

- The Multilayer Perceptron's are the most established and least complex ones.
- The Convolutional Neural Networks (CNN) significantly adjusted for processing images.
- The recurrent neural networks are utilized for successive information, for example, times series or text.

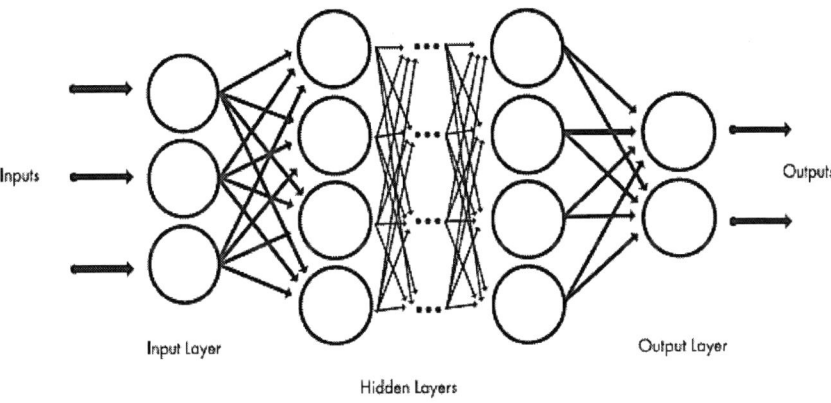

Figure 2. The architecture of Deep Neural Network.

5.2. LITERATURE SURVEY

5.2.1. Survey about FER

Archana Verma and Lokesh Kumar Sharma (2013) proposed a survey in which the objective of this work was to present an overview of the structure of analyzing facial expression. The processes engaged with expression reviews like face acquisition, feature extraction, and classification of expression had been considered. Every progression was considered with the methodologies and strategies which could be utilized to achieve the proposed objective. Facial expression recognition was the solution to future generation HCI frameworks. We have picked an increasingly incorporated methodology when contrasted with the vast majority of the general uses of FACS. Extricated facial features were utilized in a collective way to discover the best facial expression [1].

Samiksha Agrawal et al. (2015) proposed a review on FER. The unbiased view of this study was to present the current developments in FER and the associated areas in a way that needs to be acceptable even by the new scholars. To do as such, they have reviewed the different parts of FER in detail. The methods utilized in past Human Facial expression detection identified them based on computational time, accuracy, and various algorithms. Some of them contain disadvantages as far as the recognition rate or timing. The highest optimum recognition rate could be acquired through a combination of methods presented and extricate their features according to their requirement, and the last comparison was made to discover the outcomes. The achievement of performance-based was upon the pre-processing stage on the images due to illumination and extraction of the feature. Appropriate Static conditions must be connected to improve them [42].

Claude C. Chi Belushi and Fabrice Bourel (2002) have reviewed about automated expression detection. Comparative models and methods for processing were frequently utilized for FER and face identification, despite the duality which presented among these identification undertakings. 2D monochrome facial images sequence was the most prominent sort of images utilized for automated expression detection. Although an assortment of face recognition methods has been designed, reliable detection and detection of faces or their constituents were hard to accomplish in several cases. Features for automated expression recognition mean to obtain dynamic or static facial data particular to separate expression. Kinetic, geometric, and statistical or spectral-transform related features were regularly utilized as an optional description of the facial expression before classification. A broad range of classifiers, covering parametric just as nonparametric methods, has been connected to automated expression recognition [16].

W. Zhao et al. (2003) concluded a chapter where they presented speculation about face recognition dependent on psychological analysis and lessons learned from developing algorithms. They speculate that various techniques were engaged with human identification of recognizable and new faces. For instance, it was conceivable which 3D head models were developed, by expanded training for recognizable faces, however for new faces, multi-view 2D images were saved. This suggested that they have full probability density capacities for recognizable faces, while for new faces they just have discriminant functions [39].

Maja Panic and Ioannis Patras (2006) presented a framework for automated recognition of facial AUs and their temporal models from large, sequences profile-view face image. Rather than representing another way to deal with machine analysis of prototypic facial expressions of emotion, the technique introduced in this work attempted to deal with a massive scope of human facial conduct by perceiving facial muscle activities that produce expressions. Most of the present vision system for facial muscle activity identification manages just face images of frontal-view and cannot deal with temporal elements of facial motions. They utilized particle filtering to follow 15 facial points in an input face-profile sequence, and they presented facial-action elements detection from continuous video input utilizing temporal standards. The algorithm performed both automated segmentation of an input video into pictured facial expression and detection of temporal sections (i.e., offset, apex and onset) of 27 AUs that happened separately or in integration in the input face-profile video. A recognition level of 87% was accomplished [37].

Shan Li and Weighing Deng (2019) presented a detailed study on deep FER, including datasets and algorithms, which give bits of knowledge into these fundamental issues. Initially, they presented the available datasets which were generally utilized in the literature and provided accepted selection of data and assessment standards for these datasets. They described the standard pipeline of a deep FER framework with the related background information and proposals of relevant usage for every stage. For the better deep FER, they surveyed existing novel deep neural networks and related training techniques, which were intended for FER dependent on both static and dynamic image sequences and discusses their benefits and drawbacks. Competitive performances on extensively utilized benchmarks were likewise condensed in this area. Hence, they extended their review to many related issues and application situations. Finally, they reviewed the rest of the difficulties and related probabilities in this field just as future bearings for the plan of robust deep FER frameworks [7].

B. Fazel and Juergen Luettin (2003) presented the most eminent program automatic facial expression analysis and techniques are discussed in this review. Facial motion and deformation extrication techniques, just as classification techniques, were analyzed as for issues, for example, face normalization, facial expression intensity, and facial expression dynamics. Also, stated that if programmed automatic facial expression analysis frameworks were to be operated independently, current feature extraction strategies must be enhanced and increased concerning stability in natural environments just as the autonomy of manual intercession during deployment and initialization. Furthermore, a more critical analysis of some delegate facial expression analysis frameworks and examined deformation and motion-related feature extraction frameworks, hybrid systems dependent on multiple complementary face processing tasks, and multimodal systems combined, for example, acoustic and visual signals [33].

Andrea F. Abate et al. (2007) proposed this chapter, which provided an "excursus" of late face recognition analysis slants in 2D imagery and 3D model related algorithms. To simplify comparatives transverse over different procedures, tables, including a group of parameters, (for example, input measure, recognition rate, count of addressed issues) were given. This chapter finished up by proposing conceivable future bearings. Biometrics demonstrated an appropriate option but on the other hand, they experience the disadvantages also. Scanning of the iris, for example, was entirely dependable yet excessively intrusive; fingerprints were socially acknowledged, yet not adopted to non-consentient individuals. Then again, face recognition represented a decent concession among what is socially adequate and what is dependable, though while working under controlled situations. In the most recent decade, numerous algorithms dependent on nonlinear/linear techniques, neural networks, wavelets, and so forth have been presented. Generally, the Face Recognition Vendor Test 2002 demonstrated most of these methodologies experienced issues in outdoor circumstances. This reduced their dependability in contrast with best-in-class biometrics [19].

Swati Mishra and Avinashi Dhole (2013) presented this review chapter. Different FER methods and their corresponding regions have been analyzed. In the past FER framework, the performance was dissected, based on accuracy, computational time, and rate of recognition. In the more significant part of the present framework, there were some issues regarding the efficiency of recognition and recognition time prerequisites. The integration of existing systems could improve the rate of recognition of the framework, or new technique could likewise be utilized. The optimum rate of recognition relies upon the extraction of the feature stage in which pertinent features should be extricated and classified dependent on the classification technique [28].

A short assessment of the FER introduced by Rabia Jafri and Hamid R. Arabia (2009), and practical techniques were actualized in these gatherings, and a small amount of the upgrades and inconveniences of the strategies alluded to were analyzed. Face recognition methods could be comprehensively isolated into three classifications dependent on the face data acquisition approach: strategies that work on intensity images, those which manage video sequences and those which require other sensory information, for example, 3D data or infra-red imagery. Besides, a review laying out the incentive for utilizing face recognition, the utilization of this innovation, and a portion of the challenges provoking current frameworks as to this assignment have additionally been presented. This chapter additionally refers to the latest algorithms designed for this reason and endeavors to produce a concept of the best in the class of face recognition innovation [26].

5.2.2. Survey on Classifiers Based FER

Mandala et al. (2019) presented a review chapter and attempted to address most exploiting facial features like partial occlusion, aging, pose invariance, and illuminations. They were viewed as irreplaceable factors in the face recognition framework when acknowledged over facial images. This chapter likewise analyzed the best-in-class face recognition methods, techniques, viz. Eigenface, ANN, SVM, PCA, ICA, Elastic Bunch Graph Matching, Gabor Wavelets, Hidden Markov Models, and 3D morphable Model. Additionally, the previously mentioned works have referred to various testing face databases that comprised AT and T (ORL), FERET, AR, LFW, Yale, and YTF, individually for results analysis. Nonetheless, the point of this exploration was to present a detailed survey of face recognition in addition to its applications [52].

Asia Khanam et al. (2008) discusses that FER was of significant significance in human-computer interaction (HCI) frameworks. They introduced a new idea for FER out of facial characteristics applying the Mamdani-type fuzzy framework. Another feature of this framework was the membership function model of expression output that depends on various psychological surveys and studies. The verification of the model was additionally helped by the high level of expression recognition [47].

Nail Perveen et al. (2012) proposed a decision tree-based methodology for expression detection. This was the enhancement of the manual facial characteristic points as the process of feature extraction automatically computed the facial points through which facial parameters of animation were determined to detect various expressions. Facial expression data, for the most part, focused on facial expression data areas so the eyes, eyebrows, and mouth were extricated out of the input image; at the point when a face image was input. The feature extraction performs that support in recognizing facial characteristics. Facial parameters of animation were determined to detect one of the six fundamental facial expressions. The proposed method was implemented to the JAFFE database comprising 30 images, all having six original facial expression images. For specific expressions like the happy, fear, and surprise, it produced successful outcomes with the best detection rate, anyway for expressions like anger, sadness, and neutral, the detection rate was lower [44].

Catherine Solidi et al. (2012) proposed a novel strategy to perform precise FER by changing the appearance space into an expression space. The expression space was processed from the individual independent association of the facial expression discovered from the information. The forecast decreased the element of the demonstration space on a manifold compliant with the organization of the expression. Test results on 14 different mixed expressions demonstrated that the proposed association-based strategy enhanced the FER performance contrasted with appearance-based techniques by 13% [23].

Keith Anderson and Peter W. McCowan, (2006) presented the results for a completely automatic real-time expression recognition framework ready to recognize between expressions of happiness, sadness, surprise, disgust, fear, and anger. Faces were detected utilizing an algorithm spatial ratio layout tracker. The optical flow of the face was, in this way, decided to utilize a real-time performance of a robust gradient model. The expression recognition framework at that point averages facial velocity data over recognized areas of the face and cancelled rigid head movement by taking proportions of this averaged motion. The motion labels generated were then classified, utilizing SVM as either non-expressive or six fundamental expressions. The finished framework was shown in two simple affective computing applications that react in real-time to the facial emotions of the user, consequently giving the possibility for enhancement in the collaboration among a PC user and technology [24].

Sujata G. Bhele and V. H. Mankar (2012) proposed a review to analyze the critical number of chapters to cover the ongoing research in face recognition. This comprised PCA, LDA, ICA, SVM, Gabor wavelet soft computing systems like ANN for recognition, and different hybrid integration methods. This survey analyzed each one of these techniques with parameters which challenges face recognition like pose variation, illumination, and facial expressions. The present analysis uncovered that for upgraded face recognition, the new algorithm needs to develop hybrid strategies for soft computing mechanisms. For example, ANN, SVM, and SOM may yield better execution [25].

Roja Ghasemi and Maryam Ahmady (2014) proposed a new framework dependent on Fuzzy logic for the FER. Fuzzy was one practical methodology for classification, which could decide the internal division in a group of unlabeled information and discover representatives for homogenous groups. This framework perceived seven fundamental facial expressions, incredibly happy, surprise, fear, anger, calm, sadness, and disgust. Initially, they presented a novel technique for facial area extraction out of the static image. For the finding of practical face, regions were utilized from essential projection curves. This technique has a high capacity in the smart choice of regions in the FER framework. Extricated facial features sustained to a fuzzy logic-based framework for FER. Outcomes of tests directed on the JAFFE database demonstrated which proposed technique for FER was reliable with high precision and producing better outcomes as correlated with different methodologies [43].

Ketki R. Kulkarni and Sahebrao B. Bagal (2015) presented a comparative analysis of automated FER by balancing the impact of age on the identification procedure by Weighted Least Square filtering. The system utilizes a Gabor filter and Log Gabor filter to extricate features of the face. The SVM classifier was initially trained utilizing public input image and, after that, classified obscure input image. From exploratory outcomes, it tends to be presumed that accuracy of recognition enhanced with the utilization of Log Gabor filter though the time needed for processing was more when Log Gabor filter was utilized and correlated with Gabor filter [8].

Ashish Lonare and Shweta V. Jain (2013) proposed a review representing the Facial Expression Analysis for Emotion Recognition. This review additionally managed brief subtleties of different methodologies like optical flow technique, Pyramid of the histogram of gradient (PHOG), Facial action coding system (FACS), local binary patterns, and Local phase quantization (LPQ) strategy. The Block LBP histogram features extricate global and local features of face image yielding in higher precision. Nonetheless, the LBP technique was constrained to only frontal image classification. The accuracy was close to 97% when utilizing the actual images. Optical flow was performed in the areas of original parts during frames of the sequence of images and was classified into six raw emotions. This technique has high accuracy in correlation with different strategies and no necessities to choose landmarks manually at initial stage. For capturing shape data, features of PHOG were utilized, and for expression, lately, proposed features of LPQ were utilized. This technique had performed on the GEMEP-FERA dataset and demonstrated preferable outcomes over standard strategies. FACS coding was the fundamental strategy that evaluated the AU and classified it based on this AU. The expression was hence classified by considering this single AU or with other AU's combination [27].

Dhwani Mehta et al. (2019) proposed a brief review of the different methodologies and techniques utilized for those ways to deal with recognize human facial expressions for emotion detection. Besides, a detailed analysis of the integration of steps associated with an ML-based methodology and geometric-based methodology for face identification and emotion identification alongside classification was represented. However, revealing this overview, a comparison of accuracies was made for the databases, which were utilized as datasets for training and testing. Various types of databases were depicted in detail to give a concise framework of how the datasets were made, regardless of whether they were spontaneous or posed, dynamic or static, tested in non-lab conditions, and how different the datasets were. The conclusion from this study of databases was that RGB databases do not have the intensity labels, making it less helpful for the analyses to be performed and, henceforth, settles on the efficiency. The downsides of thermal databases were that it does not work with a variation of the pose, temperature variation, aging, and various scaling (e.g., identical twin issues). Disguises cannot be caught if the individual has put on glasses. Thermal images have a low resolution, which made an impact on the database quality. The 3D databases were not accessible in abundance to perform analysis and enhance accuracy. The accuracies of various algorithms with these databases are additionally discussed, which demonstrated that there was the chance for development in the field of emotion recognition regarding the accuracy and for identifying slight micro-expressions [40].

Muthamilselvan and S. Karthikeyan (2019) discusses that Face recognition was a challenging assignment in the field of analysis of the image and computer vision, which has obtained a colossal way of consideration in the route of the few decades due to its numerous applications in large fields. Many conventional face recognition strategies were referred to in this work. In a few face databases, the techniques for SVM and HMM could generate better face recognition results. However, they utilize increasingly tricky algorithms. Investigation has been focused disproportionately on this arena, and huge progression has been proficient, superior outcomes have been attained, and present face acknowledgment agendas have risen to a specific proportion of development when forced under compelled circumstances. In any case, these techniques were a long way from accomplishing the perfection of having the option to perform enough in all the different circumstances which were generally confronted by the applications utilizing these techniques in real life [22].

Jaimini Suthar and Narendra Limbad, (2014) presented the analysis and execution of the FER system that improved the Performance of FER and decreasing difficulty and represented the General FER module comprising of Pre-processing, Feature Extraction, and classification. From the evaluated chapters, there are two techniques utilized for feature extraction and various classifiers for expression classification. There was a hybrid methodology of Edge detection, extraction of features, and accurate classifier utilized dependent on a database and features and parameters of facial images. This hybrid approach is having a decent recognition rate correlated to different techniques for FER, and execution was high because of fragment the facial image into interesting expression areas like eye, eyebrows, and mouth, which are small contrasted with the total image. For extraction of feature, Gabor wavelet features having an exceptional rate of recognition. Along these, they presumed that work using a hybrid approach, Gabor Wavelet, and a facial part might expand the performance of FER [2].

Brais Martinez et al. (2014) presented a detailed survey of the review into the machine analysis of facial actions. They methodically analyzed all processes of such frameworks: pre-processing, feature extraction, and machine coding of facial actions. Likewise, the current FACS-coded facial databases are discussed. Finally, challenges that must be routed to make automated facial actions analysis appropriated to all practical circumstances are widely discussed. Automatic FACS coding would make this analysis quicker and more generally appropriate, opening new ways to understand how they convey through facial emotions. Such an automatic procedure could likewise increment the temporal resolution of coding, accuracy, and reliability [32].

Somia Saeed et al. (2006) discussed these intricate FER methods and fundamentally dissect them. In this chapter, the performance of different strategies, such as AERS, GSNMF algorithm, TPTSR, execution-based character animation, temporal template technique, feature extractions utilizing Gabor filter, and image sequencing-based techniques have been analyzed regarding their accuracy, efficiency, and viability. The accuracy and efficiency of the methods have been correlated utilizing different benchmarks like cross-validation, leave one out, and receiver operating characteristics. Every method carries its very own points of interest and detriments as far as efficiency and accuracy are concerned. The maximum rate of accuracy is shown by the method utilizing the chamfer image technique and canny edge detection algorithm [31].

Vandana Patidar et al. (2018) analyzes the various kinds of FER methods and different techniques that were utilized by them and their performance measures. This chapter provides an overview dependent on the timeline view, which performed a review on various face identification, extraction of features, and techniques of classification to deal with facial expression to detect the face. It was seen that improvement of an automatic framework which achieved FER with high accuracy of classification for constrained sorts of datasets under uncontrolled situations (like posed, occlusion, expression variations, and illuminations). In any case, advanced methodologies (maximum state feature extraction) that included ML statistical strategies enhanced the accuracy and performance of the recognition framework. Advanced techniques like Dynamic Bayesian Network, K-NN, and Hidden Markov model have prevailed to accomplish over 85% precision [5].

Guoying Zhao and Matti Pietikäinen (2007) proposed this chapter in which region-related local descriptors were utilized to perceive facial expressions in sequences of the image, integrating data from the region, volume, and region levels. The accuracy of recognition at 96.26% was accomplished on the Cohn-Kanade database of facial expression, surpassing the outcomes of the initial analysis. Assessment over a range of resolutions of the image and frame rates demonstrated which of their techniques overcome the best, making their methodology successful for real-time recognitions in unfortunate image acquisition situations. Their technique has a place with nonparametric techniques, implying that no suppositions about the necessary appropriations were required. It could manage spatiotemporal changes, and the evaluation was linear. A comparative methodology could likewise be utilized for perceiving other dynamic events like similarities of faces from the video [38].

Yogish Naik (2014) discusses that biometrics has now obtained more consideration. Face biometrics were valuable for an individual's validation as it is done easily and as a non-intrusive technique that perceived the face in the complex multidimensional visual system and built up a computational system for it. Initially, they presented a detailed overview of face recognition and discusses the technique and its performance. Then they represented the latest face recognition methods posting their disadvantages and advantages. A few procedures indicated here likewise improved the effectiveness of face recognition under different illumination and expression state of face images. Eigenface, Neural Networks, Fisherfaces, Elastic bunch graph matching, Dynamic Time Warping (DTW), Long Short-Term Memory (LSTM), and Geometrical feature matching are techniques discussed in this work [36].

Sumathi et al. (2012) reviewed different strategies which were analysed to recognize the facial expression. The process engaged with expression analysis like face acquisition, feature extraction, and classification of expression had been analyzed. The work likewise discusses the facial parameterization utilizing FACS activity units and the techniques which perceived the AU parameters utilizing facial expression information, which were extracted. Different sorts of facial expressions were available in the human face that could be recognized dependent on their geometric features, appearance features, and hybrid features. The two fundamental ideas of feature extraction depended on facial distortion and facial action. The expression recognition dependent on FACS and indirect or direct interpretations are additionally discussing with a portion of the ongoing research work. While numerous authors have been researching facial expression, fundamental emotions like a smile, surprise, sadness, and disgusting had been an intriguing point that has been generally analyzed. This article additionally recognized the strategies dependent on the attributes of expressions and classified the practical techniques which could be actualized. Subjects like Expression recognition while the intensity of expressions, spontaneous movement, temporal segmentation, combination of facial action elements detection, and pain analysis were still few subjects of interest that need to be analyzed [6].

P. Rajeswari and M.G. Sumithra, (2015) proposed four pre-processing strategies for illumination normalization in the images of the face: they were, the Gamma Intensity Correction technique (GIC), the Discrete Cosine Transform technique (DCT), the Logarithm Transform technique, and the Histogram Equalization technique (HE). It could generally enhance the image, and like this, the performance of facial features identification, contrasted with a non-pre-processed image. A progressively difficult technique as HE additionally provided better outcomes in a feature identification framework, even though the face image was not sensible. The techniques have been tried for images with a similar head posture and facial emotions [29].

T. Fang et al. (2015) overviewed the present works in 3D FER, which several have demonstrated successful outcomes in particular test circumstances. Be that as it may, most of the top-performing techniques yet need manual interpretation on the datasets. The strength of these techniques against landmark localization error has not yet been examined. Moreover, among the overviewed frameworks, just a group of them work with current 3D information and the recognition of AUs rather than six fundamental emotion classes. They expected that more exertion would be coordinated to this field since the related databases were accessible. The fact is none of the presented methodologies addressed to the difficulties of unconstrained expressions likewise calls for specialized databases.

Furthermore, the real-time reaction was usually an ideal element for any HCI framework. Henceforth, how to diminish the evaluation difficulty of 3D FER was one more fascinating issue. Finally, they proposed to build up a pair of standardized protocols with the goal that reasonable correlations could be drawn between the test results [20].

Dinesh Kumar and Rosiline Jeetha (2018) proposed a review with many kinds of literature relating to FER. Numerous techniques and systems were proposed for FER, which comprised the fuzzy support vector machine (FSVM), boosted NNE (neural network ensemble), k-nearest neighbour (KNN), multi-task facial inference model (MT-FIM), transfer subspace learning approach, sparse coding algorithm, biased subspace learning approach, and Bayesian network classifiers that were analyzed. Out of all the referenced methodologies, the FSVM with the KNN system performed better for accuracy. Be that as it may, nearby happens to be a specific analysis scope that was sufficiently competent to enhance accuracy. This could be accomplished by using image processing strategies, such as feature selection, noise removal, and classification [3].

5.2.3. Survey on Feature-Based FER

Faisal Ahmed et al. (2014) presented a stable facial feature configuration developed with the Compound Local Binary Pattern (CLBP) for person-independent FER, which outperformed the confinements of LBP. The proposed operator of CLBP consolidates extra P bits with the real LBP code to build a robust feature descriptor that utilized both the sign and the extended data of the differences among the local and the center gray values. The recognition execution of the presented strategy was assessed utilizing the CK and the JAFFE database with an SVM classifier. Test outcomes with prototypic expressions demonstrated the prevalence of the CLBP feature descriptor against few notable appearance-based feature representation techniques [50].

Yongqiang Li et al. (2013) proposed a hierarchical system dependent on Dynamic Bayesian Network for synchronous facial feature tracking and FER. By methodically representing and modeling inter-relationships between various dimensions of facial actions, just as the temporal advancement data, the proposed model accomplished huge enhancement for facial feature tracking and AU detection, contrasted with best techniques. For six fundamental expressions detection, their outcome was not satisfied with that of best-in-class techniques, since they did not utilize any estimation, particularly for expression, and the global expression was linearly deduced from AU and facial feature point estimations and their associations. The enhancements for facial feature points and AUs originated, for the most part, from consolidating the facial action model with the image estimations. In particular, the failure facial feature estimations and the AU estimations could be offset by the model's included associations between various dimensions of facial actions and the temporal connections. Since their model efficiently captured and combined the earlier information with the image estimations, with enhanced image-based computer vision methodology, their framework might accomplish the best outcomes with slight modifications to the model [51].

Chongsheng Zhang et al. (2018) discussed that the primary issue in the FER was how to extricate vital facial features, which could likely distinguish various expressions, where others could use the conventional, designed feature extraction techniques like Gabor, SIFT, and LBP. In any case, lately, CNN has been demonstrated to have the option to extricate features out of low to high states automatically and overcomes the best in the accuracy of recognition in several pattern recognition performances. In this research, they proposed a new I^{2}CNN technique to deal with the adverse impact of inter-person expression changes on the recognition of a face. Contrasted with the present conventional CNN methods, their proposal was depended on images of multi-scale global and local facial patches that could fundamentally accomplish better execution on FER. At last, they validated the efficiency of their proposed system through analysis on the public benchmarking datasets extended to Cohn-Kanade (CK+) and JAFFE [48].

Michael Revina and W.R. Sam Emmanuel (2019) described the review of FER methods, which comprised of the three main stages, pre-processing, extraction of feature, and classification. This review discussed the different sorts of FER methods with their primary commitments. The execution of different FER strategies was correlated dependent on the count of expressions perceived and the intricacy of algorithms. Databases like CK, JAFFE, and some various types of facial expression databases were analyzed in this study. The analysis performance was done depends on the database, rate of complexity, the accuracy of recognition, and significant contributions. The segmentation of ROI strategy was utilized for pre-processing, and it presents the most astounding accuracy of 99%. As indicated by feature extraction, GF has less complexity that presents the accuracy dependably somewhere in the range of 82.5% and 99%. The highest accuracy of recognition was at 99%, performed by the SVM classifier, and it perceived the many emotions, for example, smile, disgusting, surprise, sadness, fear, anger, and neutral successfully. In 2D FER, for the most part, CK and JAFFE databases were utilized for efficient performance than various databases [4].

5.2.4. Survey on Deep Learning Based FER

Zhiding Yu and Cha Zhang (2015) proposed a DCNN based FER technique, with various enhanced systems to additionally improve the performance. The proposed strategy included a face recognition model dependent on the group of three best in class face detectors, trailed by a classification model with the group of multiple deep CNN. Each CNN model was started randomly and pre-trained on a large dataset given by the FER Challenge. The pre-trained models were then adjusted on the training set of SFEW 2.0. To integrate multiple CNN models, they presented two plans for learning the group weights of the network reactions: by limiting the loss of log-probability and by reducing the link loss. This advanced technique accomplished better outcomes on both SFEW and FER datasets, indicating their outward FER technique [49].

Abir Fathallah et al. (2018) proposed a novel deep neural network design for FER. They presented a new architecture network related to CNN for FER. They modified their architecture with the Visual Geometry Group (VGG) model to enhance outputs. The proposed system comprised four layers of convolutional; the initial three layers were trailed by max-pooling, and the final one was trailed by a layer wholly connected. It accepted facial images as the input and classified them into any six expressions of the face. The architecture proposed was assessed with Ck+, RAFD, and MUG databases. Outcomes and rates of recognition demonstrated which of their techniques overcome best in class techniques. For this work, they trained the model with images in which the face was in a single position. Acquired outcomes successfully demonstrated the CNN technique's image expression recognition on numerous open databases, which accomplished enhancements in facial expression analysis [46].

Byoung ChulKo (2019) presented an overview of FER techniques. Initially, conventional FER approaches were represented alongside an outline of the delegate classifications of FER frameworks and its primary algorithms. Deep learning-based FER techniques utilizing deep networks empowering "end to end" learning was then exhibited. This review likewise targeted a unique hybrid deep learning method integrating a CNN for the features of spatial of an individual casing and LSTM for temporal features of sequential frames. Moreover, a detailed overview of openly accessible computational measurements was provided, and a correlation with benchmark results, which were a standard for a quantitative correlation of FER research, was portrayed. This chapter additionally presented some public databases associated with FER comprising of both sequences of video and still images. In a conventional data set, human facial expressions have been considered utilizing either static 2D images or 2D video sequences. Be that as it may, the fact that a 2D-based review has issues processing considerable differences in posture and unobtrusive facial actions, present datasets have considered 3D facial emotions to support more significant an analysis of the delicate structural modification's characteristic to unconstrained emotions [21].

Siyue Xie and Haifeng Hu (2019) proposed a novel technique called DCMA-CNN, a solution for the FER issue. The proposed technique was a deep based structure that was majorly comprised of two parts of CNN. One branch extricates local features out of image patches while the others extricate holistic features out of the entire expressional image. In this model, local features represented expressional subtleties, and comprehensive features described the high state semantic data of expression. They combined both local and comprehensive features in before making the classification. These two sorts of hierarchical features represented expressions in various scales. Contrasted with modern techniques with a single sort of feature, this model could describe expressions more in detail. Moreover, in the training phase, a new pooling procedure named Expressional Transformation-invariant pooling (ETI-pooling) was proposed to manage difficulties, for example, variant illuminations, rotations, and so on. The comprehensive analysis was led on the well-known JAFFE and CK+ expression datasets, where the recognition results acquired by this model were better than most existing FER techniques [45].

Wisal Hashim Abdulsalam et al. (2019) described the background of facial emotion recognition and presented the related works. Some of the accessible public datasets for researchers were additionally included. An overview of a few most recent five years of articles from 2013 to 2019 demonstrated various methods utilized for feature extraction and classification, which a few specialists utilize separately; others utilize a combination of these methods to get an advantage of more than one of them. There were no combined methods described in this field. The trend is that ongoing exploration was towards utilizing DL, particularly CNN, and results achieved in their analysis were promising.

Gozde Yolcu et al. (2018) proposed a deep learning system to detect facial expressions automatically. This examination was structured as an initial phase in improving a non-invasive, objective, quantitative framework for neurological disease determination and observing with a definitive objective of enhancing the condition of care—the presented system comprised of a cascade of two structures of CNN. The first CNN structure was prepared to isolate facial segments. The second CNN was prepared to execute the classification of facial expression. The presented two-step framework permitted classification of guided image and combination of part-based and comprehensive data. Initial tests accomplished accuracy of 93.43% for facial expression detection, over 6% enhancement over FER from raw images of input [41].

Veena Mayya et al. (2017) proposed a novel technique for automatically recognizing facial expressions utilizing features of Deep Convolutional Neural Network (DCNN). The proposed model targeted on perceiving the facial expressions of a person from a single image. The time off feature extraction was altogether decreased because of the utilization of a general-purpose graphic processing unit (GPGPU). From an assessment of two freely available facial expression datasets, they have discovered that utilizing features of DCNN could accomplish the best-in-class rate of recognition. The facial features were extricated utilizing a DCNN utilizing Caffe on CUDA empowered GPU system. Since the GPU based Caffe module was utilized to perform the test, the duration required to extricate features was substantially decreased. The proposed model could be adjusted to any nonexclusive facial expression's recognition data set, including recognition in video sequences or still images. No retraining or broad pre-processing procedures were needed to utilize the proposed technique for facial feature extraction [34].

Bing-Fei Wu and Chun-Hsien Lin (2018) presented two fundamental contributions to this chapter. One contribution was that the proposed pre-processing technique could support the CNN model to obtain higher accuracy in the utilization of facial image processing. The other contribution was that three sorts of AFMs could reformulate the features of new samples that do not have label data with the goal that some misclassified samples could be adjusted that implies it could tune a standard model for adjusting to a particular condition. Weighted Center Regression Adaptive Feature Mapping (W-CR-AFM) was majorly considered to change the feature dissemination of testing samples into trained samples. Additionally, AFMs could be conveyed to real-time frameworks since it learned group by group instead of figuring most of the training and testing information in one group. The concept drift issue was controlled on account of AFMs map the features of the testing samples to a static feature appropriation. With the pre-processing and AFMs, a light CNN could overcome the best-in-class architectures. Contrasted with the contending architectures of deep learning with similar training information, this methodology demonstrated better execution [30].

5.3. Discussions

The proposed survey was reviewed with many different types of techniques, classifiers, and feature-based articles which represented the facial expression recognition. The papers surveyed are analyzed and are part of the confirmation accuracy request, and the documents appear in the methods, as seen in the table below. Different classifiers are used in that for the performance, and in which the deep learning-based classifier worked well than all the other classifiers in terms of recognition accuracy of facial expressions. The selected databases used facial expressions, and some of them used their own. As we focus on the deep learning classifiers, the advantages and disadvantages of the deep learning classifiers types are represented in the following Table 4.

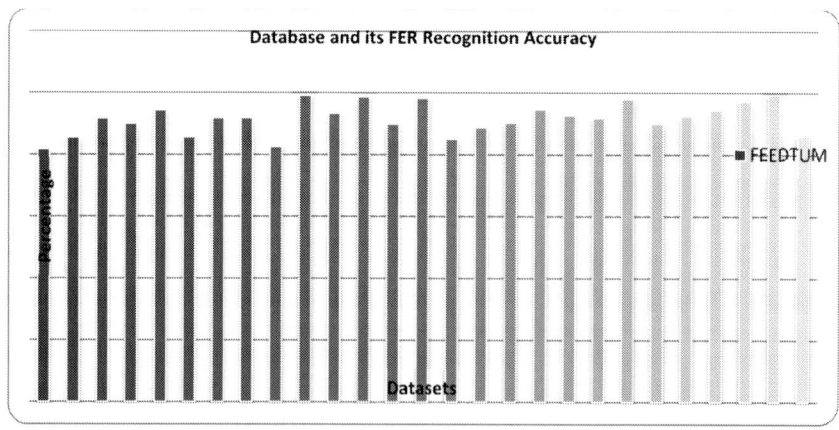

Figure 3. Comparison chart of different datasets and their accuracy.

Table 3. In-depth Study on Classifiers for Performed for FER

Article Details	Year	Classifier	Database	Recognition Accuracy
Samad et al. [70,72]	2015	Gabor Wavelet + PCA+ Multiclass SVM	FEEDTUM	81.7%
Meher et al. [71]	2014	PCA	ATT & CSU	85.5% & 81.3%
Samad et al. [70,72]	2015	Multiple Edge detection on Gabor features+ PCA + SVM	FEEDTUM	91.7%
Abdulrahman et al. [74]	2014	Gabor Wavelet + PCA + LBP	JAFFE	90%
Sobia et al. [75]	2014	PCA + FLDA	JAFFE & MUG	94.37% & 95.24%
Thai et al. [73]	2015	Canny Edge detection+ PCA + ANN	JAFFE	85.7%
Chao et al. [76]	2013	LBP + SVM	CK+	91.9%
Poursaberi et al. [77]	2012	GL Wavelet + KNN	JAFFE, CK, MMI	91.9%
Zhang et al. [78]	2014	GF + SVM	JAFFE, CK	82.5%
Hernande et al. [79]	2015	GF + SVM	KDEF	99%
Happy et al. [80]	2015	LBP + SVM	JAFFE, CK+	93.3%
Biswas [81]	2015	DCT + SVM	JAFFE, CK	98.63%
Salman et al. [82]	2017	SDM + CART	JAFFE, CK	89.9%
Kumar et al. [83]	2017	WPLBP + SVM	JAFFE, CK+, MMI	98.15%
Banu et al. [84]	2012	FFNN	Appearance-based	85%
Jizheng et al. [85]	2013	RBF	CK, BHU	88.7% & 87.8%
Jizheng et al. [86]	2013	1. SVM 2.Ada boost	JAFFE	90.3% & 94.5%
J.J. Lee et al. [87]	2008	HMM	CK	92.85%
Seyed et al. [14]	2009	GF+HFR	CK, JAFFE	91.8% & 97.9%
NaziaPerveen et al. [15]	2017	KNN	Own	90%
Liu et al. [88]	2014	3DCNN+DAP	CK+	92.4%
Liu et al. [89]	2014	STM + ExpLet	CK+	94.2%
Jung et al. [90]	2015	DTAGN	CK+, Oulu-CASIA	97.3% & 81.46%
Zhao et al. [91]	2017	PPDN	CK+, Oulu-CASIA	99.3% & 84.59%
Zhenbo Yu et al. [16]	2018	DCPN	Oulu-CASIA, CK+	86.23% & 99.6%

Table 4. Advantages and disadvantages of the Deep Learning Classifiers Algorithm Comparison table

Type of Network	Detail of Network	Advantage	Disadvantage
Deep Neural Network (DNN) [96]	There were more than two layers that allow complexion-linear relationship, and it was utilized for classification and for regression.	It was utilized widely with better accuracy	The process of training was not trivial due to the error that was propagated back to the past one layer, and they turned out very small. The process of the learning model was prolonged.
Convolutional Neural Network (CNN) [93]	This network was better for 2Ddata. It comprises of convolution filters that transform2D into 3D. It is composed of multiple building blocks.	High-performance rapid learning model, automated learning features, and adaptive learning.	For classification, it requires more labeled data.
Recurrent Neural Network (RNN) [94]	It has the capacity of sequence learning. The wights were sharing overall steps and neurons	Learn sequential events, could model time dependencies, there were few variations like HLSTM, LST, MDL-STMBLSTM. These present better accuracies in character recognition, speech recognition, and many other NLP related tasks	There may issue because of gradient vanishing and exploding gradient, the requirement of large datasets, computational nature is very slow
Deep Conventional Extreme Learning Machine (DC - ELM) [95]	For a sampling of local connections, this network utilizes Gaussian probability function	The analysis takes place on handwritten recognition is better than ELM, LRF-ELM	It is based on specific application dependent.
Deep Boltzmann Machine (DBM) [97, 98]	This model was related to the family of Boltzmann, and it has unidirectional connections among every hidden layer	The top-down feedback includes ambiguous data for a more robust analysis.	The optimization of parameters was not possible for large datasets.
Deep Belief Network (DBN) [98]	This model has a unidirectional connection between two layers on the top of layers. It was utilized in supervised and unsupervised learning in	The greedy strategy used in each layer and the tractable interference maximize the probability directly.	The initialization makes the process of training expensive by computational.

Type of Network	Detail of Network	Advantage	Disadvantage
	ML. The hidden layers of every sub-network serve as a visible layer for the next layer.		
DeepAuto Encoder [99]	It is utilized in unsupervised learning, and it is developed only for extraction and dimensionality features deduction. The amount of input was like the total output.	It does not require labeled data. There were different changes like a sparse autoencoder, De-noising autoencoder, Conventional Auto Encoder for more stability.	It needs an extra training step. Its training might suffer from vanishing.

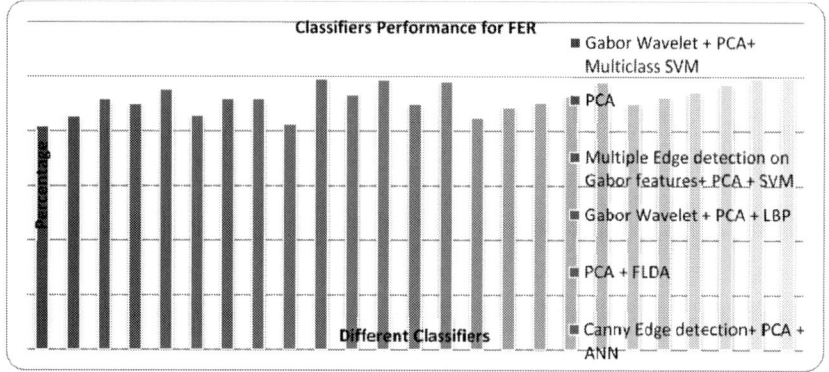

Figure 4. Comparison chart of different deep learning-based classifiers and their accuracy.

5.3.1. Conclusion

This survey has reviewed the Facial Expression Recognition techniques and the different classifiers used to classify the images collected from different databases like CK+, JAFEE, CK, MUG, and deep learning-based classifiers techniques. Moreover, several types of classifiers were used to classify facial expressions based on image processing and feature extraction techniques by other researchers were also discussed in this chapter by employing different aspects of the comparison. The comparison clearly states and shows the accuracy rate of different classifies concerning different datasets from above fig 3 and fig4. Finally, based on the depth study, it is noticed that SVM, PCA, KNN, ANN, CNN, etc. were utilized as classifiers in many types of research, especially in facial expression recognition techniques because of accuracy. Furthermore, many hybrid classifiers and more than one classifier technique were also considered for classification to accomplish better recognition and accuracy concerning the data and applications were clear from this survey. This survey was focused only on the performance of the classifier for analyzing facial expression recognition, and classification among different datasets was clearly explained and referred with suitable reference. The primary unbiased objective of this chapter is to provide the summarized result about profound learning-based classifiers benefits, advantages, and overall accuracy rate among the other techniques as well as providing future research directions for other potential reviewers was justified and summarized clearly.

CONCLUSION

This chapter reviews the concepts of facial expression recognition techniques and their categories of classifier models used for classifying facial expressions. Most operations and performances made on facial expression recognition were based on the classifiers of neural networks, machine learning, and deep learning. The deep learning-based classifier represented the best results in terms of recognition accuracy, which represents an accuracy of 99.6% for facial expressions.

Funding

No funding was applied for this work.

Competing Interests

The authors have declared that no competing interests exist.

ACKNOWLEDGMENT

We would like to express our gratitude to Jazan University and Saudi Electronic University for providing a platform for our research. Also, I would like to express our appreciation to all the previous reviewers for this chapter. Without going through their publications, we could not produce this chapter.

REFERENCES

[1] Archana V, Lokesh Kumar S. A Comprehensive Survey on Human Facial Expression Detection. *International Journal of Image Processing* (IJIP) 2013. Volume (7): Issue (2); pp.171-182.

[2] Jaimini S, Narendra L. A Literature Survey on Facial Expression Recognition techniques using Appearance-based features. *International Journal of Computer Trends and Technology* (IJCTT) - Nov 2014. Volume 17, Number 4; pp.161-165.

[3] Dinesh Kumar P, RosilineJeetha B. A Survey on Facial Expression Recognition. *International Journal of Computer Science and Engineering (IJCSE)* Dec - Jan 2018; Vol. 6, Issue 1, pp.43-50.

[4] Michael Revina I, Sam Emmanuel W.R. A Survey on Human Face Expression Recognition Techniques. *Journal of King Saud University – Computer and Information Sciences,* Elsevier 2019.

[5] Vandana P, Devang P, Dharmesh T. An Analysis of Facial Expression Recognition Techniques. *International Journal on Future Revolution in Computer Science & Communication Engineering-* Nov 2018; Volume: 3 Issue: 11; pp. 510 – 519.

[6] Sumathi C P, Santhanam T, Mahadevi M. Automatic Facial Expression Analysis A Survey, *International Journal of Computer Science & Engineering Survey* (IJCSES)- Dec-2012; Vol.3; No.6; pp-47-59.

[7] Shan L, Weihong D. *Deep Facial Expression Recognition: A Survey, Computer Vision, and Pattern Recognition-* Oct 2019. arXiv:1804.08348 [cs.CV].

[8] Ketki R. K, Sahebrao B. Facial Expression Recognition. *International Conference on Information Processing* (ICIP)- Pune, 2015, pp. 535-539.

[9] Jyothi S N, Preeti G, ManishaV, Manisha R, Samiksha S. Facial Expression Recognition: A Literature Survey. *International Journal of Computer Trends and Technology* (IJCTT)- June 2018, Volume 48; Number 1; pp.1-4.

[10] Sanjay K, Ayushi G. *Facial Expression Recognition: A Review. Special Conference Issue: National Conference on Cloud Computing & Big Data,* Shanghai, China, IJANA-2015; pp. 4-6.

[11] Ting Z. Facial Expression Recognition Based on Deep Learning: A Survey. Advances in Intelligent Systems and Interactive Applications, *Advances in Intelligent Systems and Computing* 686, Springer-2019; pp. 345-352.

[12] Sajid A K, Ayyaz H, Muhammad U. Facial Expression Recognition on Real-World Face images using Intelligent Techniques: A Survey. Optik - *International Journal for Light and Electron Optics*- April-2017.

[13] Yingli T, Takeo K, Jeffrey F. *Facial Expression Recognition.* Chapter 19, 2015.

[14] Seyed M, Zahir M. Feature Extraction for Facial Expression Recognition Based on Hybrid Face Regions, *Advances in Electrical and Computer Engineering*- 2009, Volume 9, Number 3.

[15] Nazia P, Nazir A, Abdul M, Rizwan K, Salman Q. Facial Expression Recognition Through Machine Learning. *International Journal of Scientific & Technology Research*- March 2017, Volume 5, Issue 03; pp. 91-97.

[16] Zhenbo Y, Qinshan L, Guangcan L. *Deeper cascaded peak-piloted network for weak expression recognition.* Germany: Springer-2018.

[17] Andrea F, Michele N, Daniel R, Gabriele S. 2D, and 3D face recognition: A survey. Pattern Recognition Letters 28-2007; *Science Direct*; Elsevier; pp. 1885–1906.

[18] Fang T, Zhao X, Ocegueda O, Shah S.K, Kakadiaris I.A. 3D Facial Expression Recognition: A Perspective on Promises and Challenges. *Face and Gesture*- 2015; USA; IEEE-2015.

[19] Byoung C K. A Brief Review of Facial Emotion Recognition based on Visual Information. *Sensors*, 2019, 18, 401. Available from: www.mdpi.com/journal/sensors.
[20] Tamilselvi M, Karthikeyan S. A Literature Survey in Face Recognition Techniques. *International Journal of Pure and Applied Mathematics*-2019; Volume 118; No.16; pp. 831-849.
[21] Catherine S, Nicolas S, Renaud S. *A New Invariant Representation of Facial Expressions: Definition and Application to Blended Expression Recognition.* ICIP-2012; IEEE; pp. 2617-2620.
[22] Keith A, Peter W. A Real-Time Automated System for the Recognition of Human Facial Expressions. *IEEE Transactions on Systems, Man, And Cybernetics*-Part B: Cybernetics- Feb-2006; Vol. 36; No. 1; pp. 96-105.
[23] Sujata G, Mankar V.H. A Review Chapter on Face Recognition Techniques. *International Journal of Advanced Research in Computer Engineering & Technology* (IJARCET)- October 2012; Volume 1, Issue 8; pp. 339-346.
[24] Rabia J, Hamid R. A Survey of Face Recognition Techniques. *Journal of Information Processing Systems*- June 2009, Vol.5, No.2, pp.41-67.
[25] Ashish L, Shweta V. A Survey on Facial Expression Analysis for Emotion Recognition. *International Journal of Advanced Research in Computer and Communication Engineering*- December 2013; Vol. 2, Issue 12, pp.4647-4650.
[26] Swati M, Avinash D. A Survey on Facial Expression Recognition Techniques. *International Journal of Science and Research* (IJSR)- April 2015; Volume 4, Issue 4, pp. 1247-1250.
[27] Rajeswari P, Sumithra M.G. A Survey: Pre-Processing techniques for Facial Expression Recognition. *International Journal on Applications of Information and Communication Engineering*- January 2015; Volume 1: Issue 1: Pages: 47-51.
[28] Bing F W, Chun-Hsien L. Adaptive Feature Mapping for Customizing Deep Learning-Based Facial Expression Recognition Model. *IEEE Access*- Feb 2019; pp. 12451 – 12461.

[29] Sonia S, Khalid M, Yaser D K. An exposition of facial expression recognition techniques. *Neural Comput & Applic*- Aug 2017; Springer.

[30] Brais M, Michel F, Bihan J, Maja P. Automatic Analysis of Facial Actions: A Survey. *IEEE Transactions on Affective Computing*- Sep 2014; Vol. 13, No. 9.

[31] Fasel B, Juergen L. *Automatic facial expression analysis: a survey. Pattern Recognition* 36 (2003); pp. 259-275; Elsevier- 2003.

[32] Veena M, Radhika M. P, ManoharaP.Automatic Facial Expression Recognition Using DCNN. *6th International Conference on Advances in Computing & Communications,* ICACC 2017, 6-8; September 2017, Cochin, India, Elsevier- 2017.

[33] Gozde Y, Ismail O, Serap K et al. Deep Learning-based Facial Expression Recognition for Monitoring Neurological Disorders. *2018 IEEE International Conference on Bioinformatics and Biomedicine* (BIBM); pp. 1652-1657.

[34] Yogish N. Detailed Survey of Different Face Recognition Approaches. *International Journal of Computer Science and Mobile Computing*- May 2014; IJCSMC, Vol. 3, Issue. 5; pp- 1306 – 1313.

[35] Maja P, Ioannis P. Dynamics of Facial Expression: Recognition of Facial Actions and Their Temporal Segments from Face Profile Image Sequences. *IEEE Transactions on Systems, Man, And Cybernetics-Part B: Cybernetics*- April 2006; Vol. 36, No. 2; pp- 433- 449.

[36] Guoying Z, Matti P. *Experiments with Facial Expression Recognition Using Spatiotemporal Local Binary Patterns*. ICME 2007; IEEE; pp- 1091- 1094.

[37] Zhao W, Chellappa R, Phillips P J, Rosenfeld A. Face Recognition: A Literature Survey. *ACM Computing Surveys*- December 2003; Vol. 35, No. 4; pp. 399–458.

[38] Dhwani M, Mohammad F, Ahmad Y. Facial Emotion Recognition: A Survey and Real-World User Experiences in Mixed Reality. *Sensors* 2019, 18, 416; Available from: www.mdpi.com/journal/sensors.

[39] Wisal H, Rafah S, Mohammed N. Facial Emotion Recognition: A Survey. *International Journal of Advanced Research in Computer Engineering & Technology* (IJARCET)- Nov 2019; Volume 7, Issue 11, pp. 771-779.
[40] Samiksha A, Pallavi K, Shashikant G. Facial Expression Recognition Techniques: A Survey. *International Journal of Advances in Electronics and Computer Science-* Jan 2015; Volume-2, Issue-1; pp. 61-66.
[41] Roja G, Maryam A. Facial Expression Recognition Using Facial Effective Areas, and Fuzzy Logic. *2014 Iranian Conference on Intelligent Systems (ICIS)-* April 2014; IEEE; pp. 1-4.
[42] Nazil P, Shubhrata G, Kesari V. Facial Expression Recognition Using Facial Characteristic Points and Gini Index. *Students Conference on Engineering and Systems-* May 2012; IEEE; pp. 1-6.
[43] Sigue X, Haifeng H. Facial Expression Recognition Using Hierarchical Features with Deep Comprehensive Multi-Patches Aggregation Convolutional Neural Networks. IEEE-2019; pp. 1-6.
[44] Abir F, Lotfi A, Ali D. Facial Expression Recognition via Deep Learning. 2018 *IEEE/ACS 14th International Conference on Computer Systems and Applications-* 2018; IEEE; pp. 745-750.
[45] Assia K, Zubair S M, Usman A M. Fuzzy Based Facial Expression Recognition. *2008 Congress on Image and Signal Processing*; IEEE; pp. 598-602.
[46] Chongsheng Z, Pengyou W, Ke C, Joni-Kristian K. Identity-aware convolutional neural networks for facial expression recognition. *Journal of Systems Engineering and Electronics-* August 2018; Vol. 28, No. 4; pp.784- 792.
[47] Zhiding Y, Cha Z. Image-based Static Facial Expression Recognition with Multiple Deep Network Learning. *Proceedings of the 2015 ACM on International Conference on Multimodal Interaction-* Nov 2015; ACM Digital Library; pp. 435-442.
[48] Faisal A, Hossain B, Emam H. Person-Independent Facial Expression Recognition Based on Compound Local Binary Pattern

(CLBP). *The International Arab Journal of Information Technology*- March 2014; Vol. 11, No. 2; pp. 195-203.

[49] Yongqiang L, Shangfei W, Yongping Z, Qiang J. Simultaneous Facial Feature Tracking and Facial Expression Recognition. *IEEE Transactions on Image Processing*- July 2013; Vol. 22; No. 7; pp. 2559-2573.

[50] Madan L, Kamlesh K, Rafaqat H. et al. Study of Face Recognition Techniques: A Survey. (IJACSA) *International Journal of Advanced Computer Science and Applications*- 2019; Vol. 9; No. 6; pp. 42-49.

[51] Jyoti K, Rajesh R, Pooja K M. Facial expression recognition: A survey. *Second International Symposium on Computer Vision and the Internet*- 2015; Procedia Computer Science 58; Elsevier; pp. 486 – 491.

[52] Priyanka A, Nimbarte M S. A Survey on Facial Feature Extraction to Recognize Facial Expressions. *International Journal of Engineering Research & Technology* (IJERT)- February – 2014; Vol. 3, Issue 2; pp. 539-544.

[53] Andrew R, Jeffery F, Simon L. Automated Facial Expression Recognition System. *43rd Annual 2009 International Carnahan Conference on Security Technology;* IEEE; Zurich, 2009, pp. 172-177.

[54] Monika D, Lokesh S. Automatic Emotion Recognition Using Facial Expression: A Review. *International Research Journal of Engineering and Technology* (IRJET)- Feb 2017; Volume: 03; Issue: 02; pp. 488-492.

[55] Aliaa A, Wesam A. Automatic *Facial Expression Recognition System Based on Geometric and Appearance Features, Computer, and Information Science-* March 2015; Vol. 4, No. 2; pp. 115-124.

[56] Yu-Li X, Xia M, Fan Z. Beihang University Facial Expression Database and Multiple Facial Expression Recognition. *Proceedings of the Fifth International Conference on Machine Learning and Cybernetics,* Dalian, 13-16 August 2006; IEEE; pp.3282- 3287.

[57] Nidhi N, Zankhana H, Samip A. Facial Expression Recognition: A Survey. *International Journal of Computer Science and Information Technologies-* 2015; Vol. 5 (1); pp. 149-152.

[58] Ayesha B, Maya I, Parag K. Facial Expression Recognition for Security. *International Journal of Modern Engineering Research* (IJMER)- July-Aug 2012; Vol.2, Issue.4; pp-1449-1453.

[59] Stan Z, Anil K. *Handbook of Face Recognition.* Chapter.8; Face Tracking and Recognition from Video; Springer- 2005.

[60] Ting W, Siyao F, Guosheng Y. *Survey of the Facial Expression Recognition Research.* Springer-Verlag Berlin Heidelberg 2012; pp. 392–402.

[61] Poornima P Radhapriya S. Survey of Automatic Facial Expression Recognition Based on Classification Schemes. *International Journal of Advanced Information in Engineering Technology* (IJAIET)- Oct 2018; Vol.4, No.10; pp.1-11.

[62] ChiehMing K, Shang-Hong L, Michel S. A Compact Deep Learning Model for Robust Facial Expression Recognition. *The IEEE Conference on Computer Vision and Pattern Recognition* (CVPR) Workshops- 2019; pp. 2121-2129.

[63] Seyed M, Zahir M. *Automatic facial expression recognition: feature extraction and selection.* SIViP; Springer- 2012; 6: pp.159–169.

[64] Olga K, Alex P. Facial Emotion Recognition using Min-Max Similarity Classifier. *2018 International Conference on Advances in Computing, Communications, and Informatics* (ICACCI), Udupi, 2018; pp. 752-758.

[65] Fuzail K. Facial Expression Recognition using Facial Landmark Detection and Feature Extraction via Neural Networks. *Computer Vision and Pattern Recognition-* Dec 2019; arXiv:1812.04510 [cs.CV].

[66] Hsi-Chieh L, Chia-Ying W, Tzu-Miao L. Facial Expression Recognition Using Image Processing Techniques and Neural Networks. *Advances in Intelligent Systems & Applications,* SIST 21; Springer- 2013; pp. 259–267.

[67] Sajid A, Hafiz S, Irfan A. *Feature Extraction Trends for Intelligent Facial Expression Recognition: A Survey. Informatica* 42- 2019; pp. 507–514.

[68] Leh L, Chih-Chang H, Hsueh-Yen L. Image processing-based emotion recognition. *2015 International Conference on System Science and Engineering; IEEE-* 2015; pp. 491-494.

[69] Mercy R, Durgadevi R. Image Processing Techniques to Recognize Facial Emotions. *International Journal of Engineering and Advanced Technology* (IJEAT)- August 2018; Volume-6 Issue-6; pp. 101-106.

[70] Samad, Rosdiyana, Hideyuki S. Extraction of the minimum number of Gabor wavelet parameters for the recognition of natural facial expressions. *Artificial Life and Robotics* 16, Springer- 2015; no. 1; pp. 21-31.

[71] Meher, Sukanya S, Pallavi M. Face recognition and facial expression identification using PCA. In *Advance Computing Conference,* 2014 IEEE International, pp. 1093-1098. IEEE, 2014.

[72] Samad, Rosdiyana, Hideyuki S. Edge-based Facial Feature Extraction Using Gabor Wavelet and Convolution Filters. *In MVA* 2015; pp.430-433.

[73] Thai, Le H, Nguyen D, Tran S. *A facial expression classification system integrates canny, principal component analysis, and artificial neural network.* arXivpreprintarXiv: 1111.4052; 2015.

[74] Abdulrahman, Muzammil, Tajuddeen R. Gwadabe, Fahad J, Alaa E. Gabor wavelet transform based facial expression recognition using PCA and LBP. In *Signal Processing and Communications Applications Conference, IEEE,* 2014; pp. 2265-2268.

[75] Sobia, Carmel M, Brindha V, Abudhahir A. Facial expression recognition using PCA based interface for a wheelchair. In *Electronics and Communication Systems, 2014 International Conference; IEEE,* 2014; pp. 1-6.

[76] Chao, Wei L, Jun-Zuo L, Jian-Jiun D, PO-Hung W. Facial expression recognition using expression-specific local binary patterns and layer denoising mechanism. Information Communi-

cations and Signal Processing, *2013 9th International Conference;* IEEE 2013; pp. 1-5.

[77] Poursaberi A, Noubari H.A, Gavrilova M, Yanushkevich S.N. Gauss–Laguerre wavelet textural feature fusion with geometrical information for facial expression identification. *EURASIP J. Image Video Process* - 2012; pp. 1–13.

[78] Zhang C, Wang P, Chen K. Identity-aware convolutional neural networks for facial expression recognition. *J. Syst. Eng. Electron-* 2014; 28, pp. 784–792.

[79] Hernandez A, Bonarini A, Escamilla E, Nakano M. A Facial Expression Recognition with Automatic Segmentation of Face Regions. *Int. Conf. Intell. Softw. Methodol. Tools, Tech-* 2015; pp. 529–540. doi: 10.1007/978-3-319-22689-7.

[80] Happy S.L, Routray A. Automatic facial expression recognition using features of salient facial patches. *IEEE Trans. Affect. Comput-* 2015; 6, pp- 1–12.

[81] Biswas S. An Efficient Expression Recognition Method using Contourlet Transform. *Int. Conf. Percept. Mach. Intell-* 2015. pp. 167–174.

[82] Salmam F.Z, Madani A, Kissi M. Facial Expression Recognition using Decision Trees. *IEEE 13th Int. Conf. Comput. Graph. Imaging Vis-* 2015; pp- 125–130.

[83] Kumar S, Bhuyan M.K, Chakraborty B.K. *Extraction of informative regions of a face for facial expression recognition. IET Comput. Vis-* 2017; 10, pp. 567–576

[84] Banu, Danciu, Boboc, Moga, Balan. A novel approach for facial expression recognition. *IEEE 10th Jubilee International Symposium on Intelligent Systems and Informatics;* IEEE-2012.

[85] Jizheng, Xia, Lijang, Yuli, Angelo. *Facial expression recognition considering differences in facial structure and texture. IET Computer Vision,* - 2013.

[86] Jizheng, Xia, Yuli, Angolo. Facial expression recognition based on t-SNE and adaboost M2. *IEEE International Conference on Green*

Computing and Communications and IEEE Internet of Things and IEEE Cyber, Physical and Social Computing-2013.

[87] Lee J, Zia U, Kim T S. Spatiotemporal human facial expression recognition using fisher independent component analysis and hidden Markov Model. *30th Annual International IEEE EMBS Conference* 2008.

[88] Liu M, Li S, Shan S, Wang R, Chen X. *Deeply Learning Deformable Facial Action Parts Model for Dynamic Expression Analysis.* Springer International Publishing, Berlin- 2014.

[89] Liu M, Shan S, Wang R, Chen X. Learning expression lets on Spatio-temporal manifold for dynamic facial expression recognition. *IEEE Conference on Computer Vision and Pattern Recognition,* 2014, pp. 1749–1756.

[90] Jung H, Lee S, Yim J, Park S. Joint fine-tuning in deep neural networks for facial expression recognition. *IEEE International Conference on Computer Vision,* 2015, pp. 2983–2991.

[91] Zhao X, Liang X, Liu L, Li T, Han Y, Vasconcelos N, Yan S. Peak-piloted deep network for facial expression recognition. *European Conference on Computer Vision,* 2017; pp. 425–442.

[92] I.M., Emmanuel, W.R.S. A Survey on Human Face Expression Recognition Techniques. *Journal of King Saud University – Computer and Information Sciences* (2018), https://doi.org/10.1016/j.jksuci.2018.09.002.

[93] Rikiya Yamashita, Mizuho Nishio, RichardKinh Gian Do, Kaori Togashi, Convolutional neural networks: an overview and application in radiology, *Insights into Imaging,* Vol 9, pages 611–629(2018), https://doi.org/10.1007/s13244-018-0639-9.

[94] www.medium.com/@purnasaigudikandula/recurrent-neural-networks-and-lstm-explained.

[95] Shan Pang and Xinyi Yang Deep Convolutional Extreme Learning Machine and Its Application in Handwritten Digit Classification, *Computational Intelligence and Neuroscience* Volume 2016, Article ID 3049632, 10 pages http://dx.doi.org/10.1155/2016/3049632.

[96] Yu Cheng, Duo Wang, Pan Zhou, Member, IEEE, and Tao Zhang, A Survey of Model Compression and Acceleration for Deep Neural Networks, *IEEE Signal Processing Magazine, Special Issue on Deep Learning for Image Understanding,* Published 2017, arXiv:1710.09282v9.

[97] Walter H. L. Pinaya, Ary Gadelha, Orla M. Doyle, Cristiano Noto, André Zugman Using deep belief network modeling to characterize differences in brain morphometry in schizophrenia, *Scientific Reports* volume 6, Article number: 38897 (2016).

[98] Athanasios Voulodimos, Nikolaos Doulamis, Anastasios Doulamis, EftychiosProtopapadakis, Deep Learning for Computer Vision: A Brief Review, *Computer Intelligence Neuroscience,* 2018; 2018: 7068349. Published online 2018 Feb1. doi: 10.1155/2018/7068349.

[99] Saira Banu, Mehata, "SIP Based VOIP Anomaly Detection Engine using DTV and ONR", *International Journal of Networking and Virtual organization,* Sep 2018, ISSN:1470- 9503.

In: Anomaly Detection
Editors: Saira Banu et al.

ISBN: 978-1-53619-264-3
© 2021 Nova Science Publishers, Inc.

Chapter 6

ANOMALY DETECTION FOR DATA AGGREGATION IN WIRELESS SENSOR NETWORKS

Beski Prabaharan[1,], Dr and Saira Banu[2,**]*

[1] Associate Professor, Department of Computer Science and Engineering, Vel Tech Rangarajan Dr. Sagunthala R&D Institute of Science and Technology, India.
[2] Professor, Department of Information Science and Engineering, HKBK College of Engineering, Bangalore, Affiliated to Visvesvaraya Technological University, India

ABSTRACT

Wireless Sensor Networks are collecting various data from different sources that are not easily accessible to humans. Collected data were aggregated and sent to the base station. Several data aggregation systems have been used on Wireless Sensor Networks (WSN) and it has been examined. WSN focused on privacy of the homomorphism encodings.

---

[*] Corresponding Author's Email: beskip@gmail.com.
[**] Corresponding Author's Email: Saira.atham@gmail.com.

There are various enhanced frameworks for the data integration systems that are used. In the decryption scheme, cluster heads (aggregators) are used directly in combination with cipher texts. Therefore, overhead transfers are minimized. Base station produces aggregated results, which are collected from individual nodes. The purpose of this paper is to resolve all the pitfalls, generated during this process. Although these data have been aggregated, the base station will extract all the sensing data. This property is referred as "recoverable." Experimental results show that overhead transmitting rate is better. In addition, the architecture has been extended and adopted for both heterogeneous and homogeneous WSN.

Keywords: wireless sensor networks, aggregation, homo-morphic, encryption, MAC, anomaly user

1. INTRODUCTION

Wireless Sensor Networks (WSN) are proving their importance in the various areas like military, hospitals, environment and emergency care. The WSN communicates information and data with the help of the various sensors working together. Each sensor in the system performs its computation by sensing and extracting the data from the target object by interacting with the sensors. Large volume of research is happening in the statistical field for the efficient use of the data collected from different sensors. The cluster heads uses the data aggregation technique to sorting out the data generated by the different classes of same network.

However, these devices reduce the problems in data collection during the overhead transmission. Besides, after capturing a cluster head, challenges are still there with the data sensing of its cluster members. Two concepts are used to solve the above issues. First, an information is encrypted during the transmission. Second, cluster heads are encrypted data directly without decryption. Based on these two theories, the most popularly used technique called Concealed Data Aggregation (CDA) has been suggested.

CDA in WSN offers end-to-end security as well-as network security. Cluster heads are able to perform additional operations on encrypted numerical data, when CDA implements privacy homomorphism (PH). It is presented that a term is called Recoverable Concealed Data Aggregation (RCDA). Even if the data are aggregated by cluster heads (aggregators), RCDA, in which all sensors are used to sense an information and try to send it to the base station. Two functionalities are briefly described here with the separate data. Initially, the base station (BS) will check the validity and integrity of sensing data. Next is the execution of any aggregation functions on the base station. In WSN, there are two schemes that are used such as homogeneous (RCDA-HOMO) and heterogeneous (RCDA-HETE).

2. EXISTING SYSTEM

In Secure Aggregation for Wireless Network [1], its efficiently detected wrongdoing nodes, Hu and Evans used a lightweight security architecture. SIA [2] and they recommended a way to stop the stealthy attack in, which the user is coerced by the intruder to accept inaccurate readings that result in incorrect aggregation values.

In paper [3], safe pattern-based data aggregation and Energy Efficient for Wireless Sensor Networks concept has been proposed. ESPDA offers connectivity between nodes with very little energy consumption.

This procedure is based on the clustering approach, and the pattern matching theory is used. Via the cluster-head, a pattern seed is sent to the sensor nodes. In response, the corresponding pattern code was sent by these nodes back to the cluster head.

A significant feature of user-behavior Analysis and System Anomaly Detection are used to detect anomalous user activities from incident log data and network traffic. Standard monitoring systems are unlikely to recognise new/unknown forms of anomalies as qualitative and semantic dimensions of the data for review are not integrated. User Activity Anomaly Detection uses user interaction characteristics to achieve a higher degree of detection in this setting. The log data used by User Activity Anomaly Detection has different dimensions. Data becomes scarce with increasing dimensions (attributes), and it becomes increasingly complex to detect anomalies.

This paper [4] provides details on Safe DAV: Secure Data Aggregation and its Sensor Networks Verification Protocol [5] and addresses the Secure Reference-Based Wireless Sensor Networks Data Aggregation Protocol are [6] dealt with CDA: Cellular Sensor Networks Hidden Data Aggregation. In this paper [7] discusses about SDAP: Sensor Networks Stable Hop-by-Hop Data Aggregation Protocol. The paper [8] describes the Secure and Effective Aggregation of Data for Wireless Sensor Networks. In paper [9], the author discusses about SEDAN technique for handling Data Aggregation in WSN. The author in paper [10] explains the credibility-based Safe Data Aggregation methods in WSN. The paper [11] describes about the secured transmission of data from one end to another end of data in WSN. The paper [12] discusses the secured Data Aggregation in WSN for Energy-Efficient and High Accuracy.

3. PROPOSED SYSTEM

Sensor nodes are data collecting devices scattered in different geographical locations. Cluster heads group the information from all the sensor nodes and forward the collected data to the Cluster Head (CH). Then the CH communicated the information to the sink.

Figure 1, represents data from different geographical areas to be collected, which is called Aggregation of Recoverable Concealed Data (RCDA). In RCDA, the data are sensed by sensors and these data are aggregated by cluster heads. With these findings, there are two methods that are proposed. First, the BS validates the sensing data consistency and reliability. Second, all aggregation functions can be performed.

Mykletun et al. proposed a concealed data aggregation scheme based on the elliptic curve ElGamal (EC-EG) cryptosystem. It consists of four procedures: key generation, encryption, aggregation and decryption. Symbols + and * denotes addition and scalar multiplication on elliptic curve points, respectively.

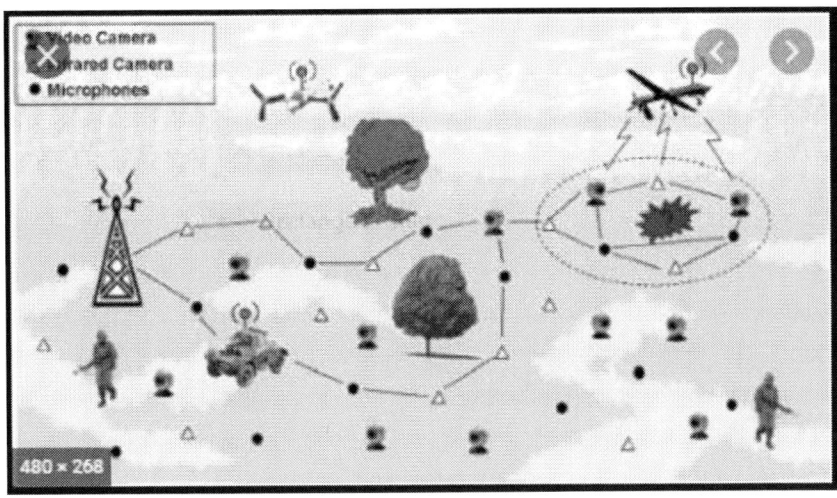

Figure 1. Wireless Sensor Nodes.

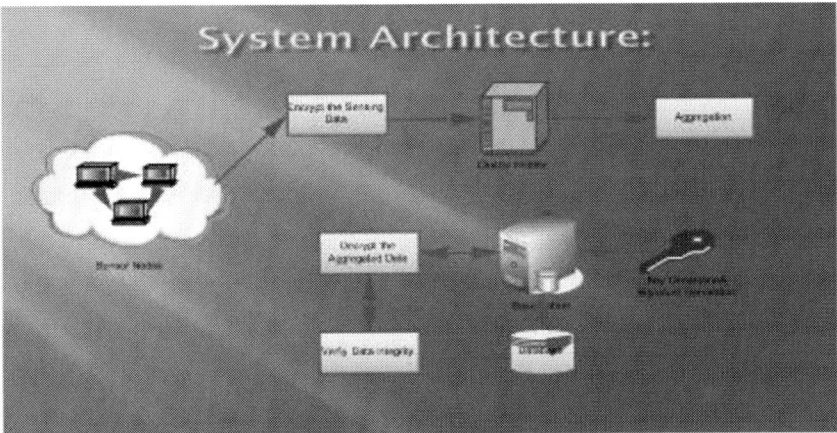

Figure 2. System Architecture.

4. METHODOLOGY

This section discusses about an implementation part. This contains the following modules that are sensor node creation, cluster header creation, base station activation, aggregation, temporal key generation & signature generation and finally verification of the integrity.

i) Mykletun et al.'s Encryption Scheme:

A hidden aggregation scheme based on the elliptic curve was suggested by Mykletun et al.

This leads to a decrease in overhead transmission and increases the average WSN lifespan.

Applications/Improvements: A new solution using homomorphic cryptography, Mykletun KeyGen and Boneh Signature Scheme is proposed to achieve secrecy, fairness and usability for safe data aggregation in ElGamal (EC-EG) cryptosystem wireless sensor networks.

It is made up of four methods:

- KeyGen: Method for generating the key
- Enc: Method for encrypting the data
- Agg: Method for aggregating the data.
- Dec: Method for decrypting the data

The symbols '+' represents addition and the symbol '*' represents the scalar multiplication respectively on elliptic curve points

Boneh's Scheme

The author Boneh [5] combines the variety of signatures into a single signature. This technique uses the following five processes:

- KeyGen: The method for generating the key
- Sign: The method for generating the signature
- Verify: The methods are used for verification
- Agg: Method for aggregating the data
- aggregation (): Method for combining the signatures
- aggregated signature verification (Agg-Verify).

Sensor Node Creation

- In this module register for number of sensor nodes.
- First, main server should be activated
- Register with name of the sensor, IP Address, and Port number.
- If already registered nodes, login with node name and port number.
- Sensor node details stored in the database.

Cluster Header

- In this module we register many cluster headers.
- Cluster Headers enter the name, IP Address, port number to register.
- Enter already registered Cluster Headers, Node name and Port Number to Login.
- Cluster headers details stored in the Database.

Base Station Activation

- In this module the base station is to be activated.
- Enter base station name, IP Address, Port number to Activate.
- Base Station details stored in the Database.
- Login Base Station to activate sensor nodes and cluster headers.

Aggregation and Temporal Key Generation and Signature Generation

- After registering details of sensor nodes, cluster header and Base station.
- Sensor node sends sensing data to Base station via cluster header.

- Cluster header receives the data and aggregates it.
- After that it generates temporal key and Signature for data.
- Encrypted data sent to the Base Station.

Verify and Integrity of the Data

- Base Station received the Encrypted Data and decrypts it.
- Verify the Veracity of the information with the key and Signature.
- Store the Data wherever it wants.

5. RESULT AND ANALYSIS

In wireless sensor networks, one of the important concepts is aggregation of the data. In this work, data can be aggregated and send it to the base station through which, energy of the network could be saved. This saved energy can be used in the network, which can improve the lifetime of the network. Results and experiments show that through with the data aggregation the performance of the WSN is improved.

Table 1. Cryptographic primitives used

Scheme	Message authentication	Digital signature	Symmetric key	Public key	Readings commitment	Privacy homomorphic	Broadcast authentication	Interactive protocol	Voting scheem
CDA			x			x			
SDA	x		x				x		
SIA	x		x		x		x	x	
SHDA	x		x		x		x	x	
RA									
WDA	x		x						x
SecureDAV		x		x	x				
SRDA			x						
SDAP	x		x		x		x	x	
ESA	x		x						
EDA			x			x			

Table 1, shows that, different methods of cryprographic primitives are used in the scheme. The authentication code is used to avoid the user to modify the messages. Public and symmetric encryptions are used to preserve the data from eavesdropping in end-to-end encryption. Additionally, verification method adds more security during the data communication.

CONCLUSION

This paper produces an implementation of recoverable hidden data aggregation for the heterogeneous and homogeneous WSN, a special benefit being that all sensing data may be efficiently obtained by the base station. The overhead transfer is still suitable when we look into an aggregated information of, yet. In addition, one can apply the signature aggregate technique to ensure data legitimacy and veracity in the design. Although signatures bear extra expenses, the suggested techniques are feasible for WSN environment. By using 100 nodes, simulations were performed in WSN environment.

REFERENCES

[1] Hu L., Evans D., "Secure Aggregation for Wireless Networks," In: *International Symposium on Applications and the Internet*, Orlando, Florida, USA, pp. 384-391, 27-31 January 2003.

[2] Przydatek B., Song D., Perrig A., "SIA: Secure Information Aggregation in Sensor Networks," In: *proceedings of the 1st International Conference on Embedded Networked Sensor Systems*, Los Angeles, CA, USA, pp. 255-265, November 05-07, 2003.

[3] Cam H. et al., "ESPDA: Energy-Efficient and Secure Pattern-Based Data Aggregation for Wireless Sensor Networks," In: *Computer*

Communications, Elsevier, Volume 29, Issue 4, pp. 446-455, February 2006.

[4] Mahimkar A., Rappaport T. S., "Secure DAV: A Secure Data Aggregation and Verification Protocol for Sensor Networks," In: *IEEE Conference on Global Telecommunications*, Volume 4, pp. 2175-2179, 29 Nov. 3 Dec. 2004.

[5] Ozgur Sanli H., Ozdemir S., Cam H., "SRDA: Secure Reference-Based Data Aggregation Protocol for Wireless Sensor Networks," In: *IEEE 60th Conference on Vehicular Technology*, VTC2004-Fall, Volume 7, pp. 4650-4654, 26-29 September 2004.

[6] Girao J., Schneider M., Westhoff D., "CDA: Concealed Data Aggregation in Wireless Sensor Networks," In: *IEEE International Conference on Communications*, Volume 5, pp. 3044-3049, 16-20 May 2005.

[7] Yang Y. et al., "SDAP: A Secure Hop-by-Hop Data Aggregation Protocol for Sensor Networks," In: *Journal of ACM Transactions on Information and System Security (TISSEC)*, Volume 11, Issue 4, Article No. 18, New York, USA, July 2008.

[8] Ozdemir S., "Secure and Reliable Data Aggregation for Wireless Sensor Networks," In: *proceedings of 4th International Symposium*, UCS 2007, Tokyo, Japan, pp. 102-109, 25-28 November 2007.

[9] Bagaa M. et al., "SEDAN: Secure and Efficient Protocol for Data Aggregation in Wireless Sensor Networks," In: *proceedings of 32nd IEEE Conference on Local Computer Networks*, pp. 1053-1060, 15-18 October 2007.

[10] Alzaid H., Foo E., Nieto J. G., "RSDA: Reputation based Secure Data Aggregation in Wireless Sensor Networks," In: *proceedings of 9th IEEE International Conference on Parallel and Distributed Computing, Applications and Technology*, pp. 419-424, 1-4 December 2008.

[11] Poornima. A. S., Amberker B. B., "SEEDA: Secure End-to-End Data Aggregation in Wireless Sensor Networks," In: *proceedings of 7th IEEE International Conference on Wireless and Optical Communications Networks (WOCN)*, pp. 1-5, 6-8 September 2010.

[12] Li H., Lin K., Li K., "Energy-Efficient and HighAccuracy Secure Data Aggregation in Wireless Sensor Networks," In: *Journal of Computer Communications*, Elsevier, Volume 34, Issue 4, pp. 591-597, 1 April 2011.

[13] Chen C. M. et al., "RCDA: Recoverable Concealed Data Aggregation for Data Integrity in Wireless Sensor Networks," In: *IEEE Transactions on Parallel and Distributed Systems*, Volume 23, Issue 4, pp. 727-734, August 2011.

[14] Jose J., Princy M., Jose J., "PEPPDA: Power Efficient Privacy Preserving Data Aggregation for Wireless Sensor Networks," In: *IEEE International Conference on Emerging Trends in Computing, Communication and Nanotechnology*, pp. 330-336, 25-26 March 2013.

[15] Wang T., Qin X., Liu L., "An Energy-Efficient and Scalable Secure Data Aggregation for Wireless Sensor Networks," In: *International Journal of Distributed Sensor Networks*, Hindawi Publications, Article ID 843485, Volume 2013(2013).

[16] Saira Banu, Mehata, "VOIP Performance Enhancement through SPIT Detection and Blocking," *Special Issue: Emerging Technologies in Networking and Security*, IIOAB, Dec 2016, Volume 7, Page: 754-763, ISSN: 0976-3104.

[17] Saira Banu, Mehata, "SIP Based VOIP Anomaly Detection Engine using DTV and ONR," *International Journal of Networking and Virtual organization*, Sep 2018, ISSN: 1470-9503.

In: Anomaly Detection
Editors: Saira Banu et al.
ISBN: 978-1-53619-264-3
© 2021 Nova Science Publishers, Inc.

Chapter 7

ALGORITHM FOR REAL TIME ANOMALOUS USER DETECTION FROM CALL DETAIL RECORD

Saira Banu[1,], PhD and Beski Prabaharan[2,†], PhD*

[1]Associate Professor, Department of Information Science
and Engineering, Vel Tech Rangarajan Dr. Sagunthala R&D Institute
of Science and Technology, India.
[2]Associate Professor, Department of Computer Science
and Engineering, Chitkara University Institute of Engineering
and Technology, Chitkara University, Rajpura, Punjab, India

ABSTRACT

The advertisers and the telemarketers are considered as spammers by the legitimate callers. Detecting the calls from the spammers before attending is a challenging task in the telecommunication industry. This algorithm finds the anomalous users from the data available in the Call Detail Record (CDR). The CDR contains all the details of each call like,

---

[*] Corresponding Author's Email: Saira.atham@gmail.com.
[†] Corresponding Author's Email: beskip@gmail.com.

incoming call number, incoming call duration, outgoing call duration and outgoing number. The proposed algorithm uses the above-mentioned parameters for detecting the anomalous users. The existing methods detect the spammers after attending the call, but the proposed algorithm detects the spam callers before attending the call.

1. INTRODUCTION

There are still billions of people connected around the world, who depend on the normal phone calls. The important threat in the call communication is the presence of spammers among the legitimate users.

Advertiser and telemarketers are using the communication system for sending unwelcome bulk calls. The detection of spammers in real time phone calls is much difficult than the email spams because, the callee is directly connected by the incoming call. For example, a email spam that reaches the inbox at 3 a.m. is not a disturbance to the user, but a spam call at 3 a.m. will be a nuisance to the user.

2. FEATURES OF A SPAM CALLER

The following are the significant features of a spam caller:

1. The spam caller will broadcast unwanted bulk messages.
2. The spammer will produce bulk calls for the purpose of marketing and promoting the product.
3. The purpose of fraud caller is attaining financial or personal gain.
4. The spam caller makes harassment calls with the intent to threaten the caller.
5. Computer-generated unwanted calls in a programmatic way.
6. Automated Nuisance calls for the determination of marketing.

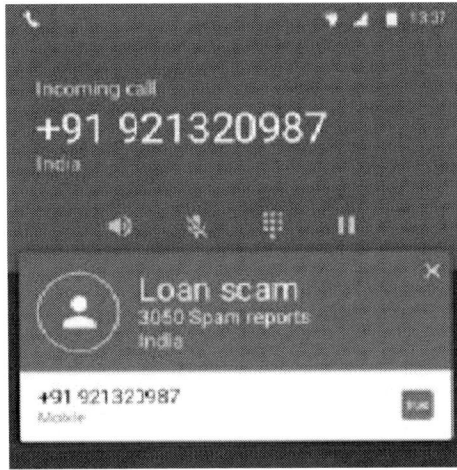

Figure 1. Spam Caller.

Truecaller App is the only android application with the capability to detect unknown caller and produce information about the unwanted call with the help of community based spam list.

The names of the customers are collected from the various phone directories globally with the help of partnership. Also, the customers registered through the Truecaller App is also listed with this data base. Appropriate algorithms are applied on the consolidated data to find the targeted name while dialing the call. The Truecaller App follows the guidelines the guidelines for maintaining the authenticity of the database. The draw back with the Truecaller App is that it cannot block before it connects the receiver. Some of the countries have enforced law against the spam callers and some countries have ratified with "do not call registry" to avoid the unwanted calls.

The image shown in Figure 1 is an example for spam call. This sender is reported as spammer by 3050 users. Google calling phone app is used to identify the suspicious call and help to avoid them before attending.

3. Survey on Spam Detection Approaches

This section surveys the various approaches used for detecting and stopping the spammers. The major categories for preventing the spammer are:

1. System-based Approach: This technique is applied for the whole internet.
2. Server-based Approach: This approach requires knowledge about the behavior of the users in the administrative domain and this approach is used in the servers.
3. Client-based Approach: This approach rejects the calls that are identified as spammers.

The other ways of categorizing the spam detection approach are

1. Content-based analysis method
2. Feedback-based analysis method [7].

The above techniques are further classified as pre and post acceptance method.

Randa et al., 2015 [16] used the supervised learning techniques for detecting the spammer with the help of behavior-based technique. Ricardo et al., 2014 [1] used the caller details for identifying the spam calls. Farideh et al., 2014 [8] have used the SIP header information for detecting the anomaly calls. Ebrahim [2] followed the ranking techniques to reduce call Spam. The parameters used for the above techniques are reputation of the caller and the feedbacks (if any) received from the called party. Dirk et al., [3] designed a detection system based on the voice data. Kentaroh et al., [4] proposed a method using the unsupervised Random Forests Classifier algorithm. This algorithm uses the Bi- directional and Incoming/outgoing Ratio for detecting the spammer. Yan Bai et al., [5] implemented the algorithm at the router level and detected the spam using the user behavior.

4. Trust Value Calculation

Dataset

The Call Detail Record contains all the information related to the caller receiver and the duration of the call. But the CDR data will not be available to the public to maintain the confidentiality of the data. The information required for this experimentation is taken from the https://uk.crawdad.org. It is a research community which provides database for experimentation. The dataset used for this experimentation has a .CSV file with more than 10000 users' information.

The dataset contains information about the call such as Customer Name, Customer Mobile Number, Routing, Call Time Duration/Sec, Call Drops, and Routing Area. Sample structure of the CDR is shown in the Figure 2.

The CDR contains the below mentioned information of each call.

- the phone number of caller party,
- the phone number of the Receiver
- the starting time of the call (date & time)
- the call duration
- the billing details of the phone number
- a unique sequence number for identifying the record
- Extra number along with the called number used for routing the call.
- Routing information through which the call reaches the exchange.
- Routing information through which the call exits the exchange.

ID	Calling Party	Called Party	Date and Time	Call Duration	Call Type	Fault Condition
123	9710410829	9003095132	24-01-2012 12.05pm	5min	Voice	Success
114	9965320235	9443799049	24-01-2012 1.30pm	7min	Voice	Success
189	9965245068	8695976414	24-01-2012 1.30pm	10min	Voice	Success
117	9597994023	9940527033	26-01-2012 03.00pm	15min	Voice	Success
145	9965216667	9094862745	27-01-2012 10.15am	8min	Voice	Success
179	9971213141	9003095132	27-01-2012 01.30pm	3min	Voice	Success

Figure 2. Structure of CDR.

$$TVC = \frac{CD_{SR}*CR_{SR}}{PO_S} + \frac{CD_{RS}}{CR_{RS}}$$

TVC is Trust value Calculation
CD is Call Duration
SR is sender to receiver
CR is call Rate
PO is the outgoing calls.

The trust value is calculated using the above formula and based on the values, the call is either accepted or rejected before reaching the receiver.

The online Reputation of the user is calculated based on the OTP (One Time Password) generated in the particular number. If an OTP reaches a mobile number, then the number is considered as a legitimate caller. The OTP increases the trust of the number.

CONCLUSION

Spammers are a major problem for the telecommunication industry. Detecting the spammers before it reaches the receiver is a major challenge. Truecaller App identifies the spam caller and displays it to the receiver. But the receiver has to see the information and then decides whether to attend or reject the call. In the proposed technique discussed in the chapter, the trust value of the caller is calculated using the information of the calls between the caller and receiver. Furthermore, OTP is also used as a main parameter for increasing the trust value of the user.

REFERENCES

[1] Ricardo Morla, Muhammad Ajmal Azad, "Caller-REP: Detecting unwanted calls with caller social strength," *Elseviar*, November 2013, Pages 219-236.

[2] Gamal A. Ebrahim, "A VoIP SPAM Reduction Framework," 2013 *IEEE*.
[3] Dirk Lentzen, Gary Grutzek, Heiko Knospe, Christoph Porschmann, "Content-based Detection and Prevention of Spam over IP Telephony - System Design, Prototype and First Results," *IEEE* 2011.
[4] Kentaroh Toyoda, Iwao Sasase, "SPIT Callers Detection with Unsupervised Random Forests Classifier," *IEEE* 2013.
[5] Yan Bail, Xiao Su, Bharat Bhargava, "Detection and Filtering Spam over Internet Telephony - A User-behavior-aware Intermediate-network-based Approach," *IEEE* 2009.
[6] He Guang-Yu, Wen Ying-You, and Zhao Hong, "SPIT Detection and Prevention Method in VoIP Environment," *IEEE,* 2008.
[7] Dongwook Shin, "Progressive Multi Gray-Levelling: A Voice Spam Protection Algorithm," *IEEE,* October 2006.
[8] Mohammad Hossein Yaghmaee Moghaddam, Mina Amanian, FaridehBarghi, and Hossein KhosraviRoshkhari, "A Survey of Different SPIT Mitigation Methods and a Presentation of a Comprehensive SPIT Detection Framework," *International Journal of Machine Learning and Computing,* Vol. 4, No. 2, April 2014.
[9] Christoph Sorge, Jan Seedorf, "A Provider-Level Reputation System for Assessing the Quality of SPIT Mitigation Algorithms," *IEEE ICC* 2009.
[10] Tetsuya Kusumoto, Eric Y. Chen, Mitsutaka Itoh, "Using Call Patterns to Detect Unwanted Communication Callers," *IEEE* 2009.
[11] Vijay A. Balasubramaniyan, Mustaque Ahamad, Haesun Park, "CallRank: Combating SPIT Using Call Duration, Social Networks and Global Reputation," CEAS 2007 *Fourth Conference on Email and Anti Spam,* August 23, 2007.
[12] Fei Wang, Yijun Mo, Benxiong Huang, "P2P-AVS: P2P Based Cooperative VoIP Spam Filtering," *IEEE* 2007.
[13] http://icsa.cs.up.ac.za/issa/2005/Proceedings/Research/091_Article.pdf.
[14] http://www.rstudio.com/products/rstudio.

[15] *CDR: Call Detail Records,* http://en.wikipedia.org/wiki/Call_detail_record.

[16] Randa Jabeue, *"Behavior Based Approach to detect Spam over IP Telephony attack,"* Springer Verlag berlin Heidelberg 2015.

[17] Saira Banu, Mehata, "SIP Based VOIP Anomaly Detection Engine using DTV and ONR," *International Journal of Networking and Virtual organization,* Sep 2018, ISSN:1470-9503.

[18] "SIP Based VOIP Anomaly Detection Engine using DTV and ONR", *International Journal of Networking and Virtual organization*, Sep 2018, ISSN:1470- 9503, Inder Science Publication, H indexed 15, Elsevier Scopus Index.

[19] "SIP Based VOIP Anomaly Detection Engine using DTV and ONR", *International Journal of Networking and Virtual organization*, Sep 2018, ISSN:1470- 9503, Inder Science Publication, H indexed 15, Elsevier Scopus Index.

In: Anomaly Detection
Editors: Saira Banu et al.

ISBN: 978-1-53619-264-3
© 2021 Nova Science Publishers, Inc.

Chapter 8

SECURED TRANSACTIONS FROM THE ANOMALY USER USING 2 WAY SSL

Syed Mustafa[1], Dr. and Mr. Madhivanan[2]

[1]Professor and Head, Department of Information Science and Engineering, HKBK College of Engineering, Bangalore, Affiliated to Visvesvaraya Technological University, India

[2]Designer Development, Natwest Group, Chennai, India

ABSTRACT

The transaction between the user and the web application needs a secured connection and trusted agreement to perform the exchange of data. This chapter describes about how the client and server application can communicate in a trusted way, using the protection of SSL technology. A trusted certificate authentication can be established by identifying both ends involved in the transactions. To perform the secured and trusted transaction, the Microservices with the help of REST API and SOAP are used using the self-assigned certificate.

Keywords: SSL, Microservice (MS), REST API, SOAP, OAuth, IAM

1. INTRODUCTION

Security is an integral part of all development projects. There could be eight design principles for securing information in computer systems [1]. To secure the transactions and information passed to the system, we have to use always HTTPS secured connection, protected password Hashing, not to expose info in URL and also to use Oauth authorization framework. The threats in transport level security in SSL/ TLS, which is used to secure the communication between sender and receiver trigger vulnerabilities in SSL/TLS by both active and passive attacks [7].

In this chapter, we use Microservices using REST API and SOAP to secure the transaction through 2 way SSL using Spring Boot Technology.

2. SSL CERTIFCATES TO PREVENT FROM ANOMALY USERS

The Microservice usage has drastically improved the application development time, speed and deployment. Software companies are developing the apps with Security architecture constructed from end-to-end security in their applications using 2-way SSL —to Microservices built using Spring Boot.

The following four methods are always used for securing the connection and transaction.

1. REST API
2. IAM authentication
3. Basic authentication using OTP
4. OAuth

3. REST API/ MICROSERVICE

Microservice is a self-container which implements a single business capability. Microservices is an architectural style that builds a web application as a set of small individual services, modeled around a business domain [2]. Here the micro services are built suing 2 –way SSL. Using a trust certificate, the server and client have to create reliance between each other. Thus, the server needs to extant a certificate to endorse itself to the client and the client has to extant its certificate to server like a digital handshake.

To perform this, first we have to create a certificate, which is self-signed by the client.

Customize the command of key tool to create the client certificate:

keytool -genkeypair -alias nt-gateway -keyalg RSA -keysize 2048 -storetype JKS -keystorent-gateway.jks -validity 3650 -ext SAN=dns: localhost,ip:127.0.0.1

Now, in the same way, we should customize the command of key tool to create the server certificate:

keytool -genkeypair -alias nt-ms -keyalg RSA -keysize 2048 -storetype JKS -keystorent-ms.jks -validity 3650 -ext SAN=dns:localhost,ip:127.0.0.1

Both client and server certificates are created and now the trust between both the client and server to be set up.

Each jks file of public certificate to be extracted to import client certificate to the server.

keytool -export -alias nt-gateway -file nt-gateway.crt -keystorent-gateway.jksEnterkeystorepassword:Certificate stored in file <nt-gateway.crt>

The diagram shows the Microservice with 2-way SSL handshake.

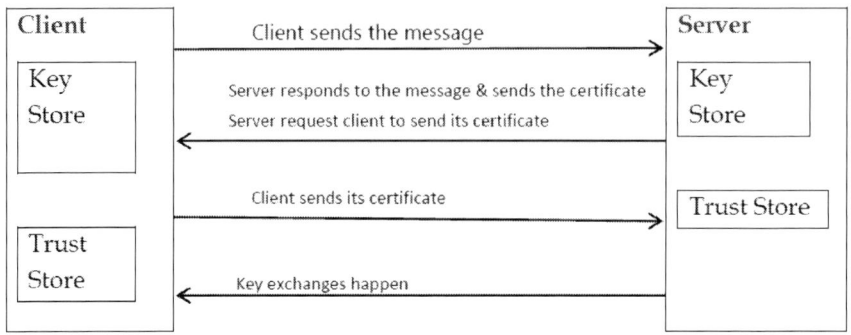

Figure 1. Microservice with 2-way SSL handshake.

We can use 2-way SSLtrusted transaction between Client/Customer and Vendor through third party payments as given in the below diagram.

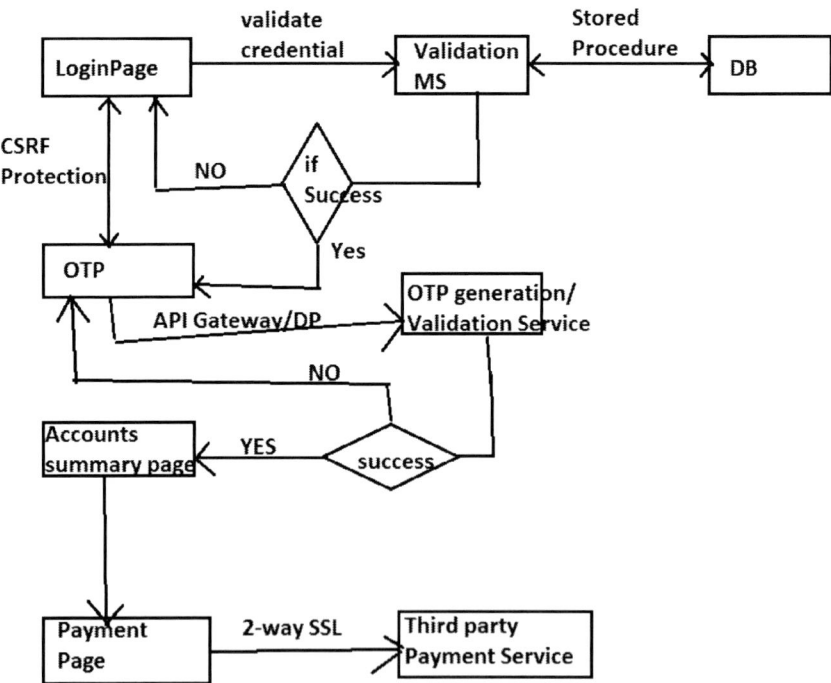

Figure 2. Card Transaction through 2-way SSL certificates.

The above diagram depicts that initially the customer logins through his or her credentials. The validation of the login credentials are verified by retrieving the details of the customer from the database through the stored procedure. If it is successful, then again, the authentication is verified through the One Time Password (OTP) to verify the customer is the right person or not by requesting the OTP generation service. If the OTP is successfully authenticated, then it goes to the account's summary page for further payment processing.

In case the OTP authentication fails, then the Customer is allowed to request for resending the OTP again. Once the customer has successfully authenticated their transaction, then the customer will land up in the payment page. Now the 2-way SSL certification is obtained from Customer to Third party payment service and from third party service to customer by sending and obtaining the trusted Certificates. Now, both the parties agree to establish the transaction so that the third party payment service like VISA, Master Card or Rupay will allow the customer to transact successfully to send and receive payments.

The below section will explain each flow with different types of security integration.

- Login page to validate Miroservice(MS)

Login page might be developed in jsp or HTML. From login page to MS we will not have any direct connectivity due to different application (if it's a JSF application which is deployed in WAS then needs no authentication between UI and backend). So basically, we use AJAX call from UI to call the microservice with the help of required header, request body and URL of Rest endpoint.

In this case, we can secure a MS with the help of OAuth token or Basic bearer or may be with client certificate by using MATLS which is transport layer security.

- Validate MS to DB

The MS will call the DB to read and validate the login credentials. In business standard, accessing the DB table with direct query will cause lot of problem so opt for with stored procedure (SP). DB is authenticated and connected with JDBC or any corresponding drivers and JNDI name and password.

The DB will return back the SP resultset, and then the result will be mapped or converted to java object. The MS will validate the username and password. If username and password are correct, then MS will return the success response.

- Login page to OTP page

Here OTP is nothing but One Time Password/Passcode. Login page to OTP page is simple static content redirection from one page to another page.

The API Gateway is nothing but redirecting or routing the client request like RCP. It usually routes/redirects the HTTP requests endpoints of the internal microservices. API Gate way will have single endpoint and it will map to multiple endpoints or group of internal microservices.

DP – Datapower is similar to API Gateway. The DataPower API Gateway is a new gateway that has been designed with APIs in mind, and with the same security focus as DataPower Gateway (v5 compatible).

- OTP validation to Summary page

Once OTP is validated, then the session is created for the logged in customer and updating cookies with session ID or any session maintaining technique. Or we can generate the OAuth token to maintain a session.

Encryption Techniques Used in Industry

1. Symmetric Encryption

Secured Transactions from the Anomaly User Using 2 Way SSL   165

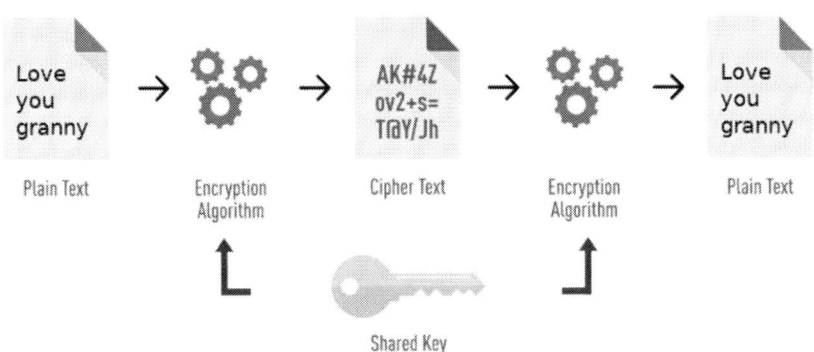

Figure 3. Symmetric Encryption.

In symmetric encryption, the same key will be used for both encryption and decryption. It only provides confidentiality, ease of use, less secured and it also requires a safe method to transfer the key from one party to another. For example, 3DES, AES, DES and RC4.

Asymmetric Encryption

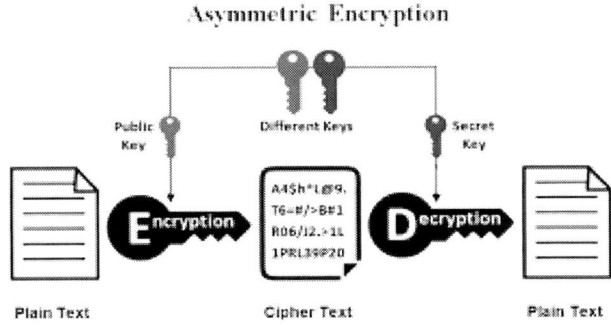

Figure 4. Asymmetric Encryption.

In Asymmetric encryption, it requires two keys, one to encrypt and the other one to decrypt and we call it as public key and private key. Certificate is the best example for extracting private key and public key.

Below commands will help to extract private key and public key from certificate (JKS – Java KeyStore). It provides confidentiality, authenticity and non-repudiation and so it is a bit slower than the symmetric algorithm. Public key and private key pair helps to encrypt information that ensures data is protected during transmission.

Public Key

Public key uses asymmetric algorithms that convert messages into an unreadable format. A person who has a public key can encrypt the message intended for a specific receiver. The receiver with the private key can only decode the message, which is encrypted by the public key. The key is available via the public accessible directory.

Private Key

The private key is a secret key that is used to decrypt the messages and the party knows it that exchange message. In the traditional method a secret key is shared within communicators to enable encryption the message, but if the key is lost, the system becomes void. To avoid this weakness, PKI (Public key infrastructure) came into force where a public key is used along with the private key. PKI enables internet users to exchange information in a secure way with the use of a public and private key. Below commands are used to extract private and public key using Openssl.

Export the Private Key from Pkcs 12 Format Keystore

- openssl pkcs12 –in keystoreName.p12 -nodes -nocerts -out private.key

Export the Public Certificate from Pkcs12 Format Keystore

- openssl pkcs12 –in keystoreName.p12 -nokeys -out public-cert-file

What Is SSL and How SSL Handshake Works

Maybe you noticed that extra "S" (http**s**) when you were browsing websites that require giving over sensitive information. In https, last character 'S' stands for SSL (Secure Socket Layer).

An SSL is security technology. It is a protocol for servers and web browsers that makes sure that data passed between the two are private. The number of packets sent or received from client to server are not affected by SSL and it is not having any impact on the network interface [6].

SSL is a double-edged sword to identify strange transactions and to decode SSL communication, technique such as SSL/TLS interception used[8].

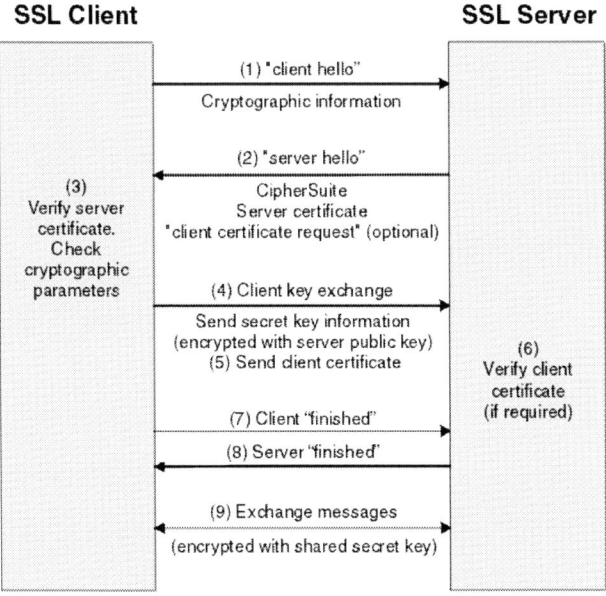

Figure 5. SSL Client -Server Communication.

4. IAM (Identity and Access Management Systems) Authentication

Authentication is needed to access any organization's network. The user will be verified with their identity to connect with the network.

Organizations may ask for additional authentication methods along with normal login credentials to provide more security such as Multifactor or Two-factor authentication (MFA or 2FA) like a key card or OTP token, or fingerprint, or a facial recognition scan. Enterprises may solve password problems by confederating user identity and covering secure single sign-on (SSO) capabilities to SaaS, cloud -based, web-based, and virtual applications by integrating password management over multiple domains, various authentications, attribute-sharing standards and protocols [3].

In this type of authentication, Client Certificate along with Private Key and PEM (Privacy-Enhanced Mail) content to generate the token. This token will be shared to the client/server for the trustful communication where malware may not be able to steal or modify the data. This IAM ping will generate the token for validation and SSO(single sign-on).

IAM provides Authorizations to determine a resources role and level of access in the network such as applications, file shares, systems, printers, etc.

The core of Identity & Access Management (IAM) manages all validation and permission processes. IAM rationalizes the entire process of managing user accounts such as Automatic User Provisioning, Workflow Management and Self-Service, Role-Based Access Control/Access Governance, Password Management, Single Sign-On (SSO, Audit & Compliance Requirements.

5. Basic Authentication

Basic authentication is built over the HTTP protocol as a simple authentication structure. The client directs HTTP requests with the

Authorization header that comprises the string 'Basic' followed by a space and a base64-encoded string username:password.

Authentication is an important process in any information system whether it is password, smart card, or biometric. Passwords are at least 12 character long and of 94 cardinality. Smart cards are composed with PIN numbers. Passwords and PIN are organized with limited attempts [4].

6. OAUTH

OAuth is created to work with Hypertext Transfer Protocol (HTTP) and allows by an authorization server to access tokens to be dispensed to third-party clients by the approval of the resource owner. To access the protected resources accommodated by the resource server, the third party uses the access token.

OAuth 2.0 lets users to share precise data with an application by keeping their usernames, passwords, and other information private. For instance, an application can use OAuth 2.0 to get permission to store files in their Google Drives from users.

REFERENCE

[1] *"The Protection of Information in Computer Systems,"* Jerome H. Saltzer, MichaelD. chroeder, http://web.mit.edu/Saltzer/www/publications/protection/.

[2] https://www.edureka.co/blog/what-is-microservices/.

[3] "Identity and Access Management: Concept, Challenges, Solutions," Mayuri Dhamdhere and Sridevi Karande, *International Journal Of Latest Trends In Engineering and Technology,* Volume 8, Issue 1, pp. 300-308, E-Issn:2278-621x.

[4] "A Review Of Authentication Methods," Nilesh A. Lal, Salendra Prasad, Mohammed Farik, *International Journal Of Scientific &*

Technology Research Volume 5, Issue 11, November 2016 ISSN 2277-8616.

[5] https://developers.google.com/identity/protocols/oauth2/web-server.

[6] "Secure Socket Layer (SSL) Impact on Web Server Performance," Mohammed A. Alnatheer, *Journal of Advances in Computer Networks,* Volume 2, No. 3, pp. 211-217 September 2014, ISSN 1793-8244.

[7] "A Comprehensive Survey on SSL/ TLS and their Vulnerabilities," Ashutosh Satapathy, Jenila Livingston L. M., *International Journal of Computer Applications,* Volume 153, No 5, November 2016, pp 31-38, ISSN 0975 – 8887.

[8] Saira Banu,Mehata, "SIP Based VOIP Anomaly Detection Engine using DTV and ONR," *International Journal of Networking and Virtual organization,* Sep 2018, ISSN:1470- 9503.

[9] "Anomaly-Based Detection of Attack on SSL Protocol using Pyod," *International Journal of Innovative Technology and Exploring Engineering (IJITEE),* Volume 9, Issue 6, April 2020, ISSN 2278-3075.

ABOUT THE EDITORS

Dr. Saira Banu received her B.E in Computer Science & Engineering from Madras University, Chennai and her M. Tech degree in Computer Science and Engineering from Anna University, Chennai. She did her doctorate in enhancing the performance of Voice using the VoIP protocol. She is currently working as Professor in the Vel Tech Rangarajan Dr. Sagunthala R&D Institute of Science and Technology She was working as a Professor in the HKBK group of Institutions, Bangalore and as an Associate Professor in the Research and Development Department of Chitkara University, Punjab. She has got 15 years of teaching experience in various universities. Her area of Research is VoIP. She has been selected as the most promising women educator of Tamil Nadu for the year 2018. She has served as a chair person and technical committee member for various international conferences. She has authored more than twenty papers in the journals, conferences and book chapter related to her research. Her research includes Wireless, Underwater IoT, Big Data Mining, Mobile Computing, Machine learning and Artificial Intelligence.

Dr. Dinesh Mavaluru obtained his PhD in Information Retrieval and currently he is working as an Educator in the Department of Information Technology at College of Computing and Informatics under the Saudi Electronic University, Riyadh, Saudi Arabia. His research interest includes

Artificial intelligence, Natural Language Processing, Data Science and Machine Learning.

Dr. Shriram Raghunathan works as the Placement Head at VIT Bhopal. He has over 20 years of experience as an educator and works on Gaming, Smart Learning and Natural Language Processing. The core mission of his work is in personalized learning and applying gamified outcomes in learning. He has executed several funded projects in Smart learning, Tamil Computing and Pervasive Computing.

Dr. Syed Mustafa obtained his Ph.D. in Computer Science and Engineering from Satyabhama University, Chennai, India. He is currently working as a Professor and the head of the Information Science and Engineering Department of HKBK College of Engineering under the VTU University. His area of research includes Web services, Web Mining, Social Media Data Mining and image Processing.

INDEX

A

access, 30, 80, 88, 102, 168, 169
adaptive handling, 2
age, 78, 91, 96, 110
aggregation, vi, viii, 67, 74, 92, 132, 139, 140, 141, 142, 143, 144, 145, 146, 147, 148, 149
algorithm, vii, 2, 5, 7, 8, 9, 12, 16, 18, 30, 32, 33, 34, 37, 38, 44, 57, 62, 86, 97, 101, 105, 109, 113, 116, 151, 154, 166
anger, 96, 108, 109, 110, 118
anomaly user, viii, 140
assessment, 105, 107, 121
authentication, 6, 86, 147, 159, 160, 163, 168
authenticity, 4, 153, 166

B

base, 20, 22, 67, 71, 78, 84, 139, 141, 143, 145, 146, 147, 153
benefits, 94, 105, 127

C

certificate, 7, 159, 161, 163, 166
challenges, viii, 17, 23, 77, 78, 79, 80, 84, 87, 88, 89, 90, 92, 107, 109, 112, 140
classes, 24, 30, 99, 100, 115, 140
classification, 24, 25, 30, 31, 32, 33, 36, 37, 38, 40, 48, 57, 69, 71, 94, 96, 98, 99, 101, 103, 104, 106, 107, 109, 110, 111, 112, 113, 114, 116, 117, 118, 120, 124, 126, 135
clustering, 24, 25, 30, 58, 59, 61, 64, 68, 74, 91, 141
clusters, 4, 48, 58, 59, 60, 61, 69
CNN, 32, 94, 99, 103, 117, 118, 119, 120, 121, 124, 126
coding, 110, 112, 116
color, 97, 98, 99
communication, 4, 49, 85, 152, 160, 168
community, vii, 19, 20, 24, 153, 155
complexity, 34, 57, 99, 118
computation, 2, 11, 49, 57, 140
computation overhead avoidance, 2
computer, 6, 11, 70, 94, 98, 101, 108, 111, 117, 160

computing, 2, 3, 11, 27, 49, 57, 78, 81, 84, 102, 109
confidentiality, 83, 155, 165, 166
consumption, 2, 4, 87
correlation(s), 25, 57, 64, 110, 116, 119
credentials, 163, 164, 168
cross-validation, 33, 62, 71, 72, 113
cryptography, 2, 10, 12, 144
customers, 81, 84, 153

D

data mining, 20, 21, 22, 25, 48
data set, 27, 34, 38, 40, 119, 121
database, 26, 100, 108, 110, 111, 112, 114, 116, 118, 145, 153, 155, 163
DDoS, 21, 78
deep learning, 32, 94, 96, 101, 102, 119, 120, 122, 126, 127
deep learning classifier, 94, 96, 122
depth, 68, 96, 123, 126
detection, vii, 14, 19, 20, 21, 22, 23, 24, 25, 30, 31, 32, 33, 34, 36, 37, 38, 39, 40, 47, 48, 51, 52, 54, 55, 56, 62, 63, 66, 68, 70, 73, 74, 94, 98, 101, 104, 105, 108, 111, 112, 113, 115, 117, 121, 123, 141, 152, 154
detection system, 21, 22, 24, 31, 36, 154
detection techniques, vii, 20, 21, 22, 25, 30, 33, 40, 70
deviation, 49, 50, 51
dimensionality, 52, 57, 66, 98, 99, 125
disgust, 96, 109, 110
distribution, 17, 54, 61, 100

E

ecosystem, 49, 77, 78
emotion, 105, 111, 115, 120, 135
encryption, 12, 83, 140, 142, 144, 147, 164, 165, 166

energy, 6, 7, 13, 16, 88, 92, 141, 146
environment(s), 2, 16, 21, 49, 55, 81, 85, 89, 90, 106, 140, 147
execution, 24, 90, 109, 112, 113, 116, 117, 118, 122, 141
extraction, vii, 19, 20, 22, 24, 27, 36, 37, 94, 96, 97, 98, 101, 103, 104, 106, 107, 108, 110, 112, 113, 114, 117, 120, 121, 125, 126, 134

F

facial expression(s), viii, 94, 96, 99, 100, 101, 103, 104, 105, 106, 108, 109, 110, 111, 113, 114, 118, 119, 120, 121, 122, 126, 127, 131, 132, 134, 135, 136, 137
false positive, 19, 21, 23, 26, 31, 32, 33, 34, 36, 40, 61
fear, 96, 108, 109, 110, 118
feature selection, 36, 40, 116
formula, 10, 11, 156

G

Germany, 47, 89, 92, 129

H

histogram, 97, 98, 110
homo-morphic, 140
human, 56, 81, 87, 91, 94, 102, 105, 108, 111, 114, 119, 137
human emotional state recognition, 94
hybrid, vii, 20, 22, 23, 31, 33, 36, 37, 38, 39, 40, 106, 109, 112, 114, 119, 126

I

IAM, 159, 160, 168
ideal, 20, 36, 57, 65, 71, 115

Index

identification, 4, 80, 96, 104, 105, 110, 111, 113, 115, 135, 136
identity, 97, 168, 170
illumination, 104, 109, 114, 115
image(s), 73, 94, 96, 97, 98, 100, 101, 102, 103, 104, 105, 107, 108, 110, 111, 112, 113, 114, 115, 116, 117, 119, 120, 121, 126, 129, 153, 172
India, 1, 131, 139, 151, 159, 172
industry/industries, 17, 92, 151, 156
integration, 70, 78, 99, 105, 107, 109, 111, 140, 163
integrity, 4, 23, 83, 141, 143
Intrusion Detection Systems, 20, 23, 34, 35, 36, 38, 41, 42, 43, 84
intrusions, 20, 21, 27, 40
IoT, vii, 2, 3, 6, 17, 43, 45, 77, 78, 79, 80, 81, 82, 84, 85, 86, 87, 88, 89, 90, 92, 171
IoT model, 78, 80, 89
IoT security, 78, 79, 82, 84, 85
issues, vii, 3, 56, 78, 79, 82, 84, 86, 89, 90, 92, 105, 106, 107, 111, 119, 140

L

lead, 2, 5, 20, 26, 56, 58, 65, 67
learning, 20, 21, 22, 30, 32, 33, 34, 52, 57, 66, 70, 73, 94, 96, 97, 101, 102, 116, 118, 119, 122, 124, 125, 127, 154, 171, 172
lifetime, 12, 14, 146
light, 4, 84, 89, 102, 121
linear model, 48, 50, 57, 58
localization, 90, 97, 115

M

MAC, 140
machine learning, 20, 21, 22, 33, 34, 52, 66, 70, 94, 127, 171
majority, 26, 52, 104

malware, 21, 42, 78, 168
mammals, 20, 22, 36
management, vii, 2, 3, 5, 17, 86, 90, 91, 168
measurements, 80, 97, 119
medical, 37, 56, 73, 90, 94
memory, 2, 4, 49, 87
messages, 4, 5, 6, 147, 152, 166
methodology, viii, 2, 5, 7, 24, 85, 99, 103, 108, 109, 111, 112, 114, 117, 122
microservice (MS), 159, 160, 161, 162, 163, 164
models, 32, 36, 47, 48, 50, 57, 59, 61, 66, 67, 68, 71, 73, 79, 89, 99, 101, 102, 104, 105, 118, 127
modifications, 97, 99, 117

N

neural network(s), 12, 16, 31, 45, 56, 57, 101, 102, 103, 105, 106, 116, 118, 127, 132, 135, 136, 137
neutral, 100, 108, 118
nodes, 2, 3, 5, 6, 12, 16, 68, 80, 84, 102, 140, 141, 142, 145, 147, 166
NSL, 31, 32, 37, 40, 43

O

OAuth, 159, 160, 163, 164, 169
operations, 11, 55, 127, 140
optimal proxies, 2
optimization, 18, 37, 40, 65, 124

P

parallel, 32, 57, 68
password, 101, 160, 164, 168, 169
PCA, vii, 37, 51, 52, 53, 54, 108, 109, 123, 126, 135
platform, 24, 32, 49, 127

population, 8, 9, 38
principal component analysis, 48, 51, 135
principles, 65, 68, 160
probability, 50, 105, 118, 124
project, 25, 30, 90
protection, 79, 82, 83, 159, 169
protocols, 4, 17, 18, 78, 79, 85, 86, 91, 92, 116, 168, 170

R

recall, 54, 55, 56, 59, 60, 61, 62, 63, 65, 67, 69, 70, 71, 72
recognition, viii, 94, 96, 98, 99, 101, 103, 104, 105, 106, 107, 108, 109, 110, 111, 112, 113, 114, 115, 116, 117, 118, 119, 120, 121, 122, 124, 126, 127, 129, 131, 132, 133, 134, 135, 136, 137, 168
regression, 48, 50, 51, 124
reliability, 5, 113, 142
reputation, 6, 89, 91, 154
requirement(s), vii, 7, 19, 20, 21, 23, 30, 33, 81, 85, 96, 104, 124
RES, 159, 160, 161
researchers, 21, 27, 33, 78, 86, 89, 94, 120, 126
resource constraints, 2, 5
resources, 6, 7, 77, 80, 84, 85, 168, 169
response, 24, 141, 164
REST API, 159, 160, 161
ROI, 97, 98, 118

S

sadness, 96, 108, 109, 110, 115, 118
Saudi Arabia, 77, 93, 171
secured key establishment, 2
security, vii, 2, 3, 5, 12, 19, 77, 78, 79, 80, 82, 83, 84, 85, 86, 87, 89, 91, 92, 140, 141, 147, 160, 163, 164, 167, 168
sensing, 92, 140, 141, 142, 145, 147

sensor(s), vii, viii, 2, 3, 4, 5, 6, 8, 17, 80, 90, 130, 131, 140, 141, 142, 143, 144, 145, 146
sensor network, viii, 3, 17, 90, 140, 144, 146
sensor nodes, 5, 8, 141, 142, 145
servers, vii, 2, 4, 5, 7, 16, 154, 167
services, viii, 6, 17, 23, 78, 81, 161, 172
shape, 58, 59, 64, 98, 99, 110
simulation(s), 2, 13, 16, 33, 34, 36, 37, 38, 147
SIP, 138, 149, 154, 158, 170
SOAP, 159, 160
social network, 77, 78, 82, 90
software, 49, 54, 82
solution, 8, 9, 10, 57, 70, 103, 119, 144
spam, viii, 101, 152, 153, 154, 156
species, 20, 22, 36, 73
speech, 101, 102, 124
SSL, vi, viii, 159, 160, 161, 162, 163, 167, 170
state(s), 20, 28, 29, 30, 40, 113, 114, 117, 120, 126
structure, vii, 4, 9, 25, 31, 32, 47, 80, 101, 103, 119, 120, 136, 155, 168
success rate, 20, 22, 26

T

target, 9, 21, 140
techniques, vii, viii, 18, 20, 21, 22, 23, 24, 25, 26, 30, 33, 36, 37, 40, 45, 48, 73, 82, 83, 84, 88, 94, 96, 98, 99, 105, 106, 107, 108, 109, 111, 112, 113, 114, 115, 116, 117, 119, 120, 122, 126, 127, 128, 130, 131, 147, 154
technology/technologies, viii, 3, 17, 49, 78, 79, 80, 86, 89, 91, 92, 109, 159, 167
temperature, 80, 81, 111
testing, 16, 24, 31, 33, 100, 108, 111, 121
threats, 4, 78, 79, 89, 160

training, 12, 24, 31, 62, 105, 111, 118, 120, 121, 124, 125
transactions, viii, 159, 160, 167
transmission, 2, 4, 5, 7, 16, 80, 85, 86, 140, 142, 144, 166
Trojans, 19, 21

U

USA, 42, 74, 88, 90, 92, 129, 147, 148

V

validation, 71, 72, 114, 163, 164, 168

variations, 97, 101, 113, 124
velocity, 8, 9, 40, 109
vision, 17, 94, 98, 101, 105, 111, 117
visualization, 10, 49, 60, 66
vulnerability, 23, 78, 81, 87

W

wavelet, 109, 112, 135, 136
web, 159, 161, 167, 168, 169, 170
wireless sensor networks, viii, 140, 144, 146
Wisconsin, 38, 64, 65, 68, 70